BRITISH CERAMIC PROCEEDINGS
NO. 56

CERAMIC OXYGEN ION CONDUCTORS AND THEIR TECHNOLOGICAL APPLICATIONS

Recent British Ceramic Proceedings

50 Engineering Ceramics:
Fabrication Science & Technology (1993)
Edited by D. P. Thompson

51 Nanoceramics (1993)
Edited by R. Freer

52 Electroceramics:
Production, Properties and Microstructures (1994)
Edited by W. E. Lee and A. Bell

53 Novel Synthesis and Processing of Ceramics (1994)
Edited by F. R. Sale

54 Ceramic Films and Coatings (1995)
Edited by W. E. Lee

55 21st Century Ceramic (1996)
Edited by D. P. Thompson

CERAMIC OXYGEN ION SEMICONDUCTORS AND THEIR APPLICATIONS

Edited by

B. C. H. Steele
Centre for Technical Ceramics
Department of Materials
Imperial College
London, UK

British Ceramic Proceedings
No. 56

THE INSTITUTE OF MATERIALS

Book 645
Published 1996 by
The Institute of Materials
1 Carlton House Terrace
London SW1Y 5DB

ISSN 0268-4373
ISBN 1-86125-003-7

The papers contained in this fifty-sixth volume
of the British Ceramic Proceedings were presented
at a meeting held at the
Lake District Conference Centre
Charlotte Mason College,
Ambleside, Cumbria, UK,
19–21 July 1995

Typeset by
Keyset Composition
Colchester, UK

Printed and bound at
The University Press
Cambridge, UK

Contents

Surface and Interface Properties of Oxides Exhibiting Fast Oxygen Transport

H.-D. WIEMHÖFER
Westfälische Wilhelms-Universität, Anorganisch-Chemisches Institut, Wilhelm-Klemm-Str. 8, D-48149 Münster, Germany

ABSTRACT
Properties of surfaces and interfaces play a central role in solid-state electrochemistry. This contribution discusses the application of surface analytical techniques to surfaces and interfaces of oxide ion conducting materials with regard to their electrochemical properties and applications. Emphasis lies on results of photoelectron spectroscopies which proved to be particularly useful. Topics are

- composition, bonding, and electronic properties of solid electrolyte surfaces,
- relation between changes of electrode potential and surface composition, surface band structure, work function, Fermi level at electrode surfaces and interfaces,
- in-situ studies for the detection of adsorbed particles and intermediates of electrode reactions to elucidate the reaction mechanisms.

Most of the presented results concern surfaces and interfaces of yttria stabilised ZrO_2 (YSZ) and electrode materials such as Pt and Ag.

1. INTRODUCTION

This contribution treats the surface and interface properties of solid oxide ion conductors with a special emphasis on their relation to the concepts of solid-state electrochemistry. This concerns, for instance, the dependence between the electrochemically controlled activities and surface quantities such as work function, Fermi level, and chemical composition. Surface analytical techniques are particularly useful for the analysis of atomistic composition, structure, electronic and other properties of surfaces. Therefore, the electrochemical concepts to be discussed in the following are largely based on results of these techniques and their application to solid oxide ion conductors. Most of the cited results concern surfaces and interfaces of stabilised zirconia and suitable electrode systems.

Table 1 shows some surface analytical techniques and typical surface properties that can be studied. UV photoelectron spectroscopy (UPS), in particular, gives a direct access to an experimental determination of work functions, surface potentials, electron affinities, and the position of Fermi levels.[1,2] X-ray photoelectron spectroscopy (XPS or ESCA) is useful in analysing the elemental composition and bonding at a surface. Rather detailed information on surface and subsurface properties can be obtained with these and some other closely related techniques based on the analysis of emitted or scattered electrons and ions.

Table 1
Examples for useful techniques for surface analysis of solid electrolytes and electrodes

Surface analytical methods	Information available
XPS (or ESCA): X-ray Photoelectron Spectroscopy (Electron Spectroscopy for Chemical Analysis)	• energy of core level electrons and valence band referred to E_F • chemical shifts due to different chemical environment • elemental surface composition • information depth: 3–10 monolayers
UPS: UV Photoelectron Spectroscopy	• surface density of occupied states in the valence band including surface states as a function of energy referred to E_F • occupied electronic states of adsorbed species • work function, ionisation energy, position of the surface Fermi level E_F with regard to the valence band edge • information depth: 2–8 monolayers
EELS: Electron Energy Loss Spectroscopy	• energy difference between occupied and unoccupied states at the surface (electronic states, surface plasmons, . . .) • vibrational states of adsorbed particles (HREELS)
AES: Auger Electron Spectroscopy	• characteristic energy differences of core levels, chemical shifts due to different chemical environment • information depth similar to XPS
SIMS/SNMS: Secondary Ion Mass Spectrometry (SNMS – secondary neutral particles)	• static: elemental composition of the first atomic layer • dynamic: depth profile of elemental composition
ISS (or LEIS): Low Energy Ion Scattering (Ion Scattering Spectroscopy)	• analysis of kinetic energy of elastically scattered ions • elemental composition of the first atomic layer (bad resolution for elements with similar mass)
TDS: Thermal Desorption Spectrometry	• mass spectrometry of thermally desorbed atoms and molecules • information obtained about adsorbed species and about number and energy of adsorption sites • detection of decomposition and changes of bulk stoichiometry

2. ELECTRODE POTENTIAL AND ELECTRONIC BAND SCHEME OF A SOLID ELECTROLYTE

With respect to surface or interface problems, one can distinguish three different parts of the electrode/electrolyte system as illustrated in Fig. 1: first, the interface between a solid electrolyte and the gas phase as well as the internal interfaces of polycrystalline electrolytes, second, the interface between electrolyte and electrode, and third, the interface between the electrode and the gas phase. These three types of interfaces are discussed in the following.

It is the Galvani potential difference at a two-phase boundary between electrolyte and electrode which characterises an equilibrium electrode potential. If both phases are good conductors, the potential difference is virtually concentrated at the interface. The consequence is that the external control of electrode potentials makes possible great changes of the potential difference and of the corresponding thermodynamic activities in a very small range at the interface. Hence, it is of interest to analyse the electrode interface as a function of electrode potential in terms of composition, atomistic and electrical properties.

A Galvani potential difference is not a measurable quantity. However, the chemical and electrical state of the electrode interface can be described within the electronic band scheme of the interface as has been pointed out earlier by Kleitz *et al.*[3,4] Figure 2 illustrates this for the example of a galvanic cell with an oxide ion conductor. The Fermi energy E_F, which is identical to the electrochemical potential

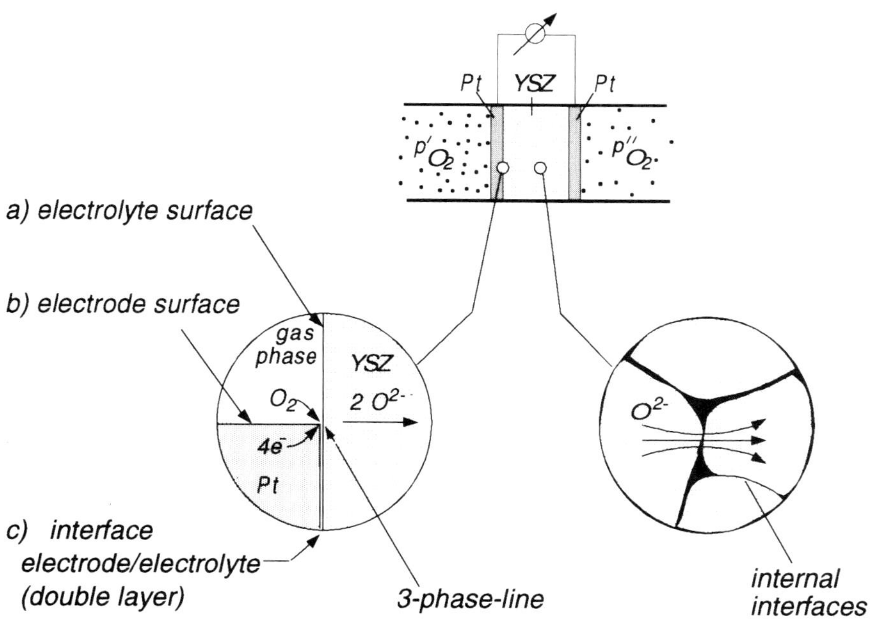

Figure 1 Relevant surfaces and interfaces in a solid-state galvanic cell with oxygen electrodes.

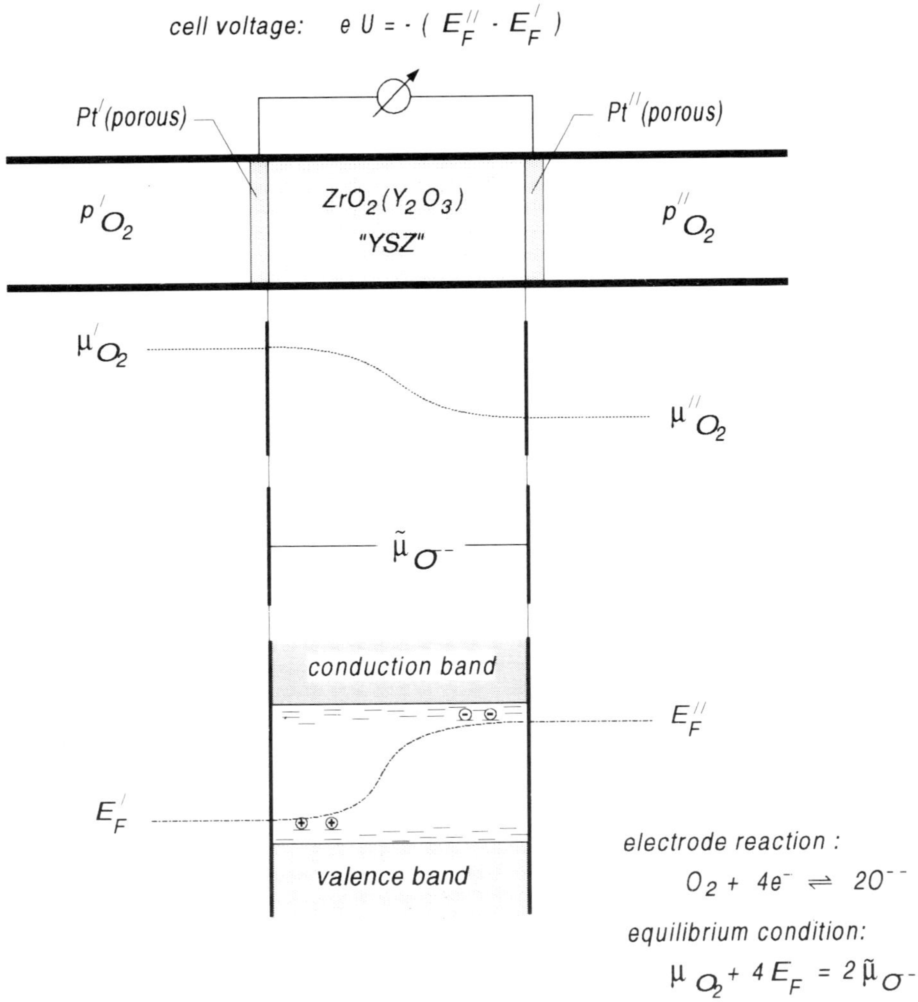

Figure 2 Meaning of cell voltage and electrode potentials in a band diagram of a solid electrolyte: the oxygen electrode reaction at Pt electrodes on stabilised zirconia (YSZ) is taken as an example (μ_{O_2} – chemical potential of oxygen, $\tilde{\mu}_{O^{2-}}$ – chemical potential of oxide ions, e – elementary charge, U – cell voltage, E_F – Fermi energy.

of electrons, plays a central role: the cell voltage corresponds to the difference of the Fermi energies between the two electrodes. Hence, there must be a Fermi level gradient within the electrolyte and a corresponding concentration gradient of electronic defects, although the concentrations of electrons and holes are usually rather small in solid electrolytes. If the electric current is zero and the ionic conductivity is high, the band bending is zero corresponding to a negligible electric field in the solid electrolyte.

The presence of gaseous oxygen or a metal oxide at the electrode interface defines a value for the oxygen activity, because then, the Fermi level is controlled at the interface by the oxygen electrode equilibrium

$$\tfrac{1}{2}O_{2(g)} + 2e^-_{(electrode)} \leftrightarrow O^{2-}_{(electrolyte)} \quad (1)$$

$$E_F = \tfrac{1}{4}\cdot(2\tilde{\mu}_{O^{2-}} - \mu_{O_2}) \quad (2)$$

In a good oxide ion conductor such as yttria stabilised zirconia (YSZ), the chemical potential of mobile oxide ions is constant due to the high vacancy concentration. In that case, the Fermi level is a definite function of the oxygen partial pressure and its change is related to the change of the corresponding electrode potential $\Delta\varepsilon$ according to

$$\Delta E_F = -\tfrac{1}{4}\cdot\Delta\mu_{O_2} = -\frac{kT}{4e}\cdot\Delta\ln p_{O_2} = -\,\mathrm{e}\Delta\varepsilon \quad (3)$$

The position of E_F as well as work function Φ, electron affinity χ and electronic state densities can be measured for a solid electrolyte surface with UPS. Hence, if the position of the Fermi level has been determined for a given electrode potential, the electrochemical electrode potential scale can be correlated to the electronic band scheme of the solid electrolyte. On the other hand, the distance between the Fermi level and the band edges or other electronic energy levels allows an estimate of the stability of an electrolyte and of the voltage limits which characterise the onset of electronic conductivity.

3. EXPERIMENTAL ASPECTS OF SURFACE ANALYSIS ON SOLID OXIDE ION CONDUCTORS

As surface analytical techniques are mostly based on excitation or analysis with charged particles, overcharging effects are to be expected, if insulating materials are measured. Although solid electrolytes are not insulating, one has to take care to use contact materials that allow the exchange of ions (and electrons). Reversible solid electrodes have to be used for that purpose. For measurements on solid oxide conductors, mixtures of binary oxides with their parent metal are favourable, for instance Fe/FeO, Cu/Cu_2O, Pd/PdO. In this way, the thermodynamic state and with it the Fermi level is unambiguously determined at the back contact of an ionically conducting sample.[5] This is a precondition for well-defined energy scales in photoelectron spectroscopies. Therefore, surfaces of solid ionic conductors can be analysed with a comparable precision as metals and semiconductors.

Concerning gas interaction at surfaces, a serious restriction of most surface analytical techniques is of course the necessity of ultra-high vacuum conditions. However, this does not exclude *in-situ* investigations of non-equilibrium gas evolution at electrodes, as Arakawa *et al.* have demonstrated in an XPS study of oxygen evolution at Ag electrodes on stabilised zirconia.[6] We have further developed this experimental technique and applied it to the measurement of electronic structure, bonding and composition of solid electrolyte and electrode surfaces as a function of temperature and electrode potential.[5,7,8,9,10]

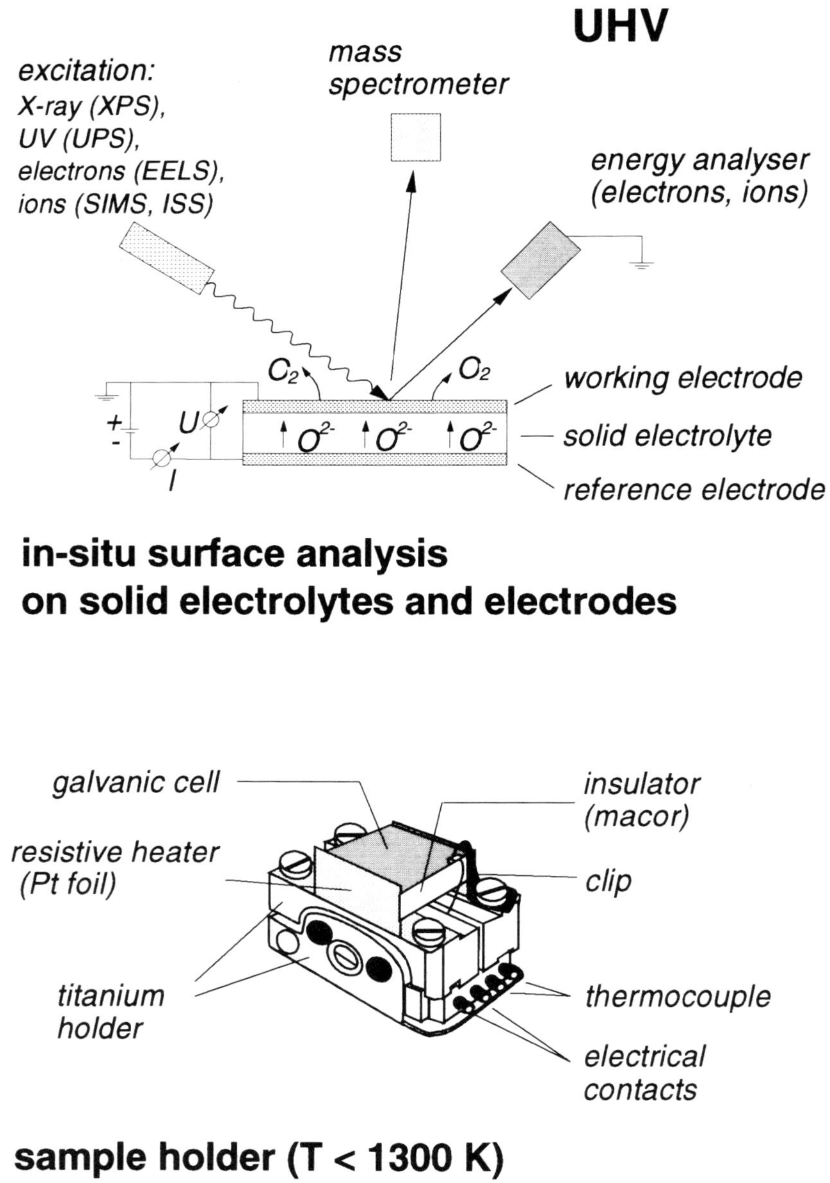

Figure 3 Experimental set-up for surface analysis of YSZ and of electrodes on YSZ

Figure 3 shows the galvanic cell employed for XPS and UPS investigations of electrodes in the temperature range between 300 K and 1250 K.[11] The cell had a layer structure. The electrolyte was ceramic or single crystal YSZ. At the back side of the electrolyte, a solid reference electrode was applied. The entire cell was mounted on a platinum foil which served as a resistive heating element. In this way, the solid electrolyte surface could be analysed without overcharging. In another series of experiments, the surfaces of various electrode materials were investigated on top of the YSZ electrolyte. The electrode potential at the front electrode was controlled with regard to the reference electrode by a potentiostat. In principle, the oxygen activity at the electrolyte/electrode interface can be varied during such experiments over many orders of magnitude of the oxygen partial pressure. Application of a constant anodic current makes possible *in-situ* studies of adsorbed species for a controlled rate of oxygen formation and in the presence of activity gradients in the electrode material. The actual partial pressure at the electrode surface depends on the kinetics. For electrodes like silver and mixed conducting oxides on zirconia, high oxygen activities were attained at the surface under UHV conditions, probably because oxygen recombination and desorption at the surface is rate-limiting.[10]

4. SURFACE PROPERTIES OF SOLID ELECTROLYTES

4.1 Surface Composition and Segregation of YSZ

In analysing a surface, one always has to bear in mind that surface, subsurface and bulk may differ in composition, lattice parameters, ionic and electronic defect concentrations and the density of electronic states. The way of surface preparation and pretreatment, as well as surface orientation and segregation of parent elements or impurities have to be considered. A defined state and composition of the solid electrolyte surface has to be achieved, before any conclusions with regard to its intrinsic properties can be made. In the following, stabilised zirconia is taken as an example.

The surface composition of the conventional ceramic electrolytes based on zirconia is complicated by the fact that segregation of a series of elements may occur. In polycrystalline ceramic materials, the main surface impurities as detected, for instance, by SIMS and ISS were Na, K, Ca, Mg, Al, Si, F and Cl. Most of them can be removed by repeated sputtering and annealing. Elements like Fe and Ti which often occur as impurities or additions in stabilised zirconia do not segregate to the surface except the case where surface ion implantation was used which led to subsurface concentrations above the solubility limit.[9,12] Nevertheless, after an appropriate cleaning and annealing procedure, the surface spectra of YSZ become reproducible and the main features of ceramic and single-crystal surfaces with the same yttria content from different sources are practically the same.

A number of groups have analysed the surface composition of stabilised zirconias by a series of surface analytical techniques.[8,13,14,15] All these investigations give clear evidence for a surface enrichment of yttria above a certain temperature. SIMS and XPS experiments on single crystals as well as on ceramic samples showed a reversible segregation occurring in a small temperature range near 500°C.[8] Typical surface concentrations found for Y by AES ranged between 22 and 34% at temperatures above 600°C.[15] This corresponds to an enrichment by about a factor of two and more.

A comparison of XPS and SIMS results (see Fig. 4) shows that yttria segregation

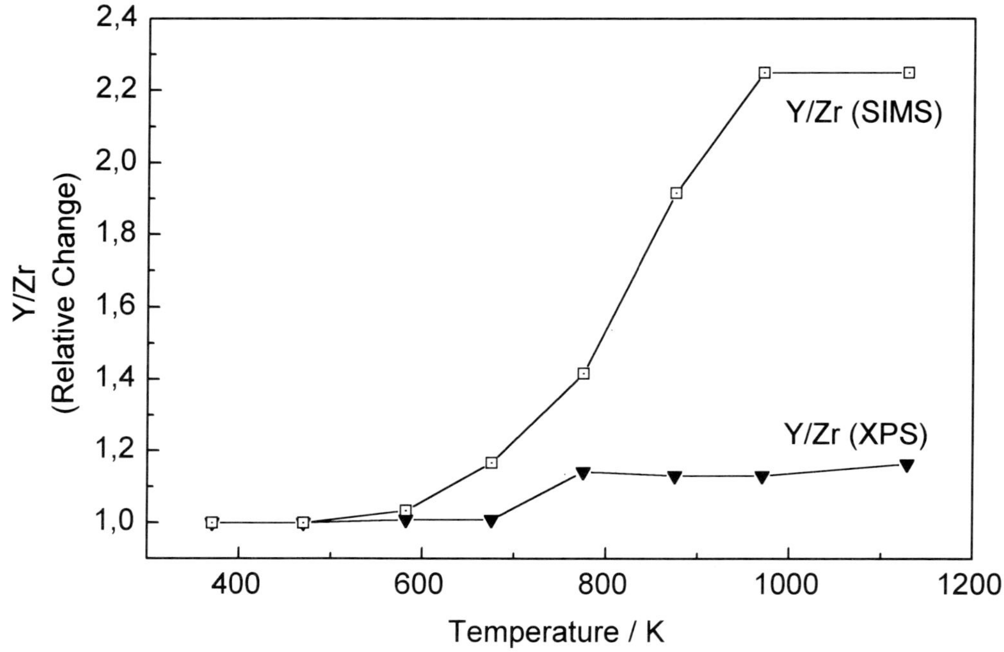

Figure 4 Results of SIMS and XPS for yttrium segregation of YSZ: Y/Zr denotes the relative quotient of (XPS or SIMS) signal intensities for Y and Zr. Its value was set equal to one at temperatures below 450 K.

is predominantly concentrated in the first monolayer. The XPS technique probes a substantially larger depth and shows a much lower yttria enrichment (only about 10%) as compared to the more surface sensitive static SIMS. This means at the same time that surface properties as determined by XPS or other techniques that also include the subsurface region will be less sensitive to effects of the yttria enrichment. Indeed, the experiments showed no indications of stepwise changes of electronic surface properties in the temperature range 500–600°C that correlated with yttria segregation. It is therefore likely that the electronic surface properties derived from UPS will not be very different from the bulk.

4.2 Spectroscopic and Thermal Band Gap of YSZ (Stabilised Zirconia)

The band gap E_{gap} and the electronic states of solid electrolytes like stabilised zirconia are of interest in order to predict or explain electron or hole conduction which is undesired in the bulk, but may be advantageous at the electrode interface. It is important to distinguish between the spectroscopic (or optical) and thermal band gap. The optical band gap E_{gap} obtained by spectroscopic techniques describes the smallest energy difference between the band edges. The thermal gap is derived from a temperature-dependent measurement of electron and hole concentrations. It then corresponds to a standard thermodynamic enthalpy ΔH^o_{eh} of formation of electron-hole pairs under equilibrium conditions (at constant pressure). If electrons and holes can be treated as quasi-free particles, the optical and thermal gap values should be nearly the same giving for the product $N_e N_h$ of electron and hole concentration

$$\frac{N_e \cdot N_h}{N_{free}^2} = \exp\left[\frac{\Delta S_{eh}^o}{k}\right] \cdot \exp\left[-\frac{\Delta H_{eh}^o}{kT}\right] = \left(\frac{m_e^* \cdot m_h^*}{m_e^2}\right)^{3/2} \cdot \exp\left[-\frac{E_{gap}}{kT}\right] \tag{4}$$

Here, $N_{free} = [2\pi m_e kT \mathrm{h}^{-2}]^{3/2}$ denotes the density of states of a free electron and is taken as the thermodynamic standard concentration. m_e^* and m_h^* are the effective masses of electrons in the conduction band and holes in the valence band. For quasi-free delocalised electrons in a band, entropy effects due to interaction with the lattice are accounted for in terms of the effective masses of electrons and holes which are used to express the deviation of the effective densities of states from that of free electrons. Then, one can set $E_{gap} = \Delta H_{eh}^{\mathrm{o}}$. For instance, if the effective masses of electrons and holes both are a factor of two larger than the mass of a free electron, the term $-T\Delta S_{eh}^{\mathrm{o}}$ for electron-hole pair formation will be of the order of 0.005 eV. This becomes negligible for optical gaps larger than 1 eV.

Electrons on localised levels, however, cannot be treated by this quasi-free electron model. An effective density of states and an effective mass is then no longer definable. The lattice surrounding the localised electron will normally be polarised with corresponding shifts of the equilibrium atomic positions. This larger interaction with the lattice (electron-phonon coupling) may give rise to a considerable vibrational entropy in addition to the energy of the electronic level. Hence, if the fractional degree of occupation of a localised electronic level has to be calculated, the difference between the free energy of the level and the Fermi energy should be used in the argument of the Fermi-Dirac function instead of only the energy of the level. Taking spectroscopically obtained energy differences as an estimate for the free energy of a localised state is only legitimate, if the vibrational entropy may be neglected.

A second source for deviations between spectroscopic and thermodynamic energies (or optical and thermal) is still more important. It is caused by the Franck–Condon principle which states that the nuclear positions remain practically constant during the time of an electronic transition. As a consequence, optical energies derived from absorption maxima tend to be higher than the thermal energy difference between the two involved electronic energy levels. The deviation corresponds to the relaxation energy of the nuclear configuration around a localised electronic defect. On the same grounds, emission frequencies are lower than absorption frequencies for a given pair of electronic states. Thus, one has to take care when discussing thermodynamic quantities derived from spectroscopic data.

Figure 5 shows the information on the (surface) band diagram that can be obtained by a combination of various optical and surface-spectroscopic techniques. The optical energy differences for transitions due to electron-hole formation and recombination are also illustrated. A number of published values exist for the band gap of stabilised zirconia which were determined by optical absorption, reflection spectroscopy and EELS (see e.g. Ref. 16 for a survey and quoted references

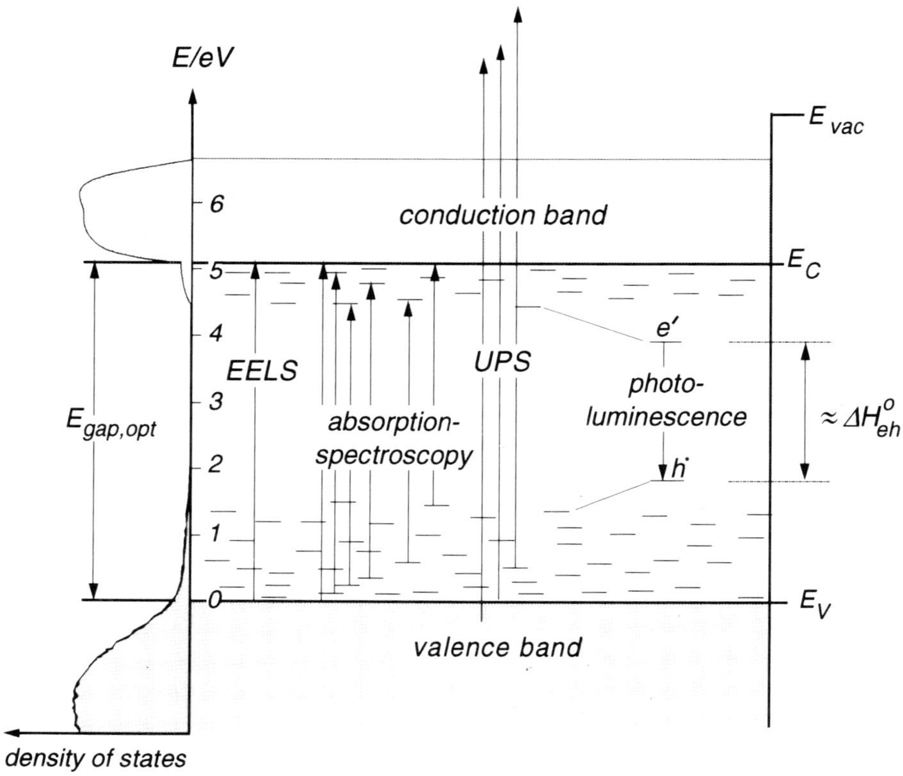

Figure 5 Determination of band edges and additional electronic levels in the band gap of YSZ by electron and optical spectroscopies (E_{vac} – vacuum level of electrons, E_C, E_V – conduction and valence band edges, E_{gap}, ΔH^o_{eh} – see text).

therein). A large scattering of the data can be observed which is largely due to differences in the interpretation of the various electronic transitions that can be measured for stabilised zirconia. The most probable reason is an appreciable density of localised electronic states in the band gap of zirconia which gives rise to absorption and emission at lower energies than the 'true band gap'. Some authors have identified these transitions with band-to-band transitions and hence got lower band gap energies near 3–4 eV. The situation is complicated by the fact that absorption bands of additional impurities are often observed depending on the kind of sample. However, the true value of the optical band gap E_{gap} (also often called optical band gap) lies near 5 eV as discussed in Ref. 16. For instance, a value of 5.2 eV was obtained with EELS.

Values for the thermal band gap of stabilised zirconia have been derived by several authors from measurements of the conductivities and diffusion coefficients of electrons and holes (see Ref. 17 and cited references therein). Park and Blumenthal, for instance, give a value of 2.61 eV for the enthalpy of formation of electron-hole pairs. The entropy of electron-hole pair formation seems to be rather high, because the same authors got an average effective density of states of 1.05×10^{22} cm^{-3} in applying the quasi-free electron model. That is three orders of magnitude higher than 3×10^{19} cm^{-3}, the approximate density of states for free electrons at 800°C. Hence, there must be a strong interaction between the lattice and the electrons or holes. The obtained electrical mobilities of electrons and holes were as low as 10^{-5} or 10^{-4} cm^2 V^{-1} s^{-1} while the concentration of electrons and holes was always higher than 10^{16} cm^{-3}.[17] It is evident that electrons and holes on localised electronic states and not conduction or valence band states cause the electronic transport in YSZ.

This is strongly supported by results of UPS and EELS, where a considerable density of electronic states was detected in the band gap of YSZ in the form of band tails extending 1–2 eV from the band edges. The density of these states was related to the defect concentration, as it increased when the ionic defect concentration was increased by sputtering.[5] It is well known that such band tails are characteristic for systems with high disorder and that they correspond to localised electronic states which are separated from the delocalised band states by a mobility edge.[18,19] It must be concluded that the thermal band gap describes the smallest distance between the band tails of the valence and the conduction band. Accordingly, the formation of localised electron-hole pairs occurs with an energy of formation that is about 2.5 eV lower than the optical band gap. It is interesting in this context that photoluminescence experiments gave energies of 2.0–2.5 eV for the recombination of electron-hole pairs in zirconia. This is the energy difference of occupied states and thus includes the interaction of the electrons and holes with the lattice. Hence, it is understandable that it corresponds nearly to the thermal band gap of 2.61 eV of Park and Blumenthal.[17] In the case of YSZ, this value is very different from the optical gap as illustrated in Fig. 5 as it corresponds to the minimum energy of localised electron-hole pair formation. It is very likely that electron or hole conduction in most solid electrolytes with their normally large band gaps involves such localised intrinsic or impurity states in the band gap.

4.3 Modification of the Band Gap by Substitution

Although it is not immediately possible to use spectroscopic energies for precise thermodynamic calculations of electron and hole concentrations, their values can be used to estimate changes of electron concentrations with electrode potentials as well as stability limits or the electrolytic domain of an ion conductor, if the position of the Fermi level in the band scheme is known. Considerable deviations are possible with respect to localised electronic states or in general when the electron–lattice interaction is strong.

The band gap of stabilised zirconia (thermal and optical) can be influenced by the addition of certain metal oxides. Dissolution of the iso-valent titanium dioxide in zirconia leads to a considerable decrease of the spectroscopic band gap by 1.6 eV.[12] A small content of titania already leads to a steep decrease of the band gap as the results of EELS in Fig. 6 show. It is assumed that the 3d states of titanium ions form a new band in a range of 2 eV below the conduction band edge of stabilized zirconia.

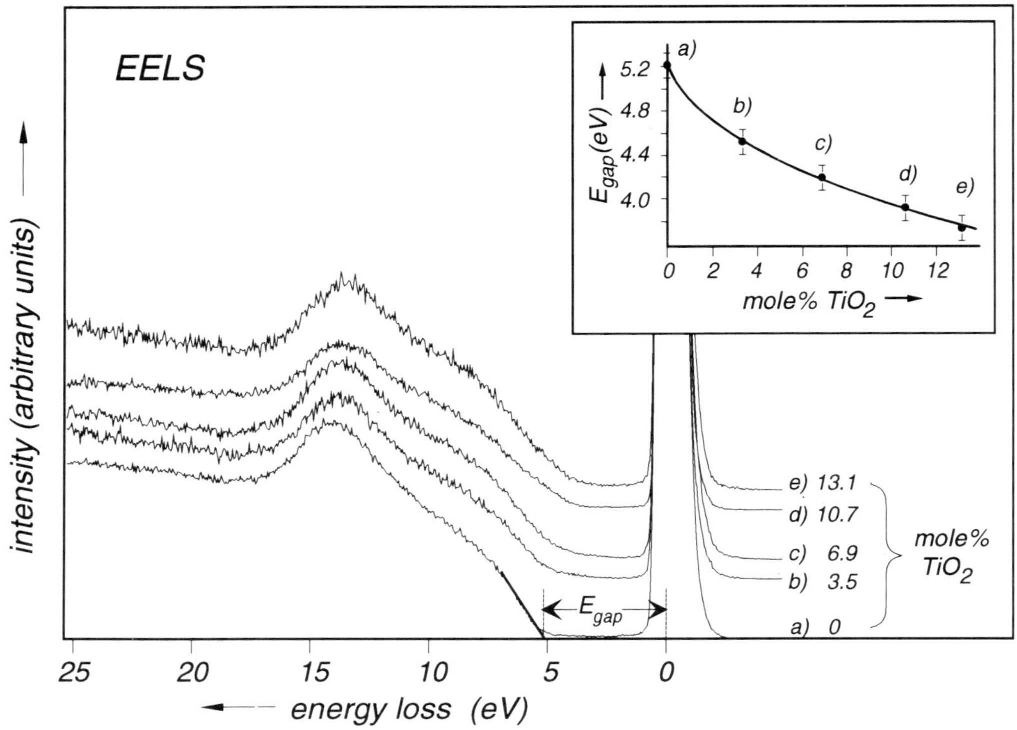

Figure 6 Spectra obtained by EELS for YSZ (10 mol% Y_2O_3) with various concentrations of dissolved titania: the inserted figure shows the lowering of the band gap of YSZ due to the addition of TiO_2.

These states shorten the energetic distance of the lowest unoccupied states to the occupied localised states above the valence band edge. Figure 7 shows the results obtained by UPS and EELS in the band diagram of YSZ. Although the band gap is greatly reduced, the position of E_F virtually remains constant with regard to the valence band edge. One can expect that the addition of titania enhances the electronic concentration under reducing conditions where the Fermi level approaches the empty 3d band of Ti^{4+} ions, which is indeed the case.[20] Similar observations have been made for stabilised zirconia with dissolved Fe_2O_3.[9] Hence, the addition of elements like Fe or Ti enhances the electronic conductivity under reducing conditions and may serve to improve the electrode kinetics under these conditions.

4.4 Work Function and Fermi Level at the Surface of YSZ

The work function is the sum of the electron affinity and the difference between Fermi energy and conduction band edge. But only the position of the Fermi level within the band scheme determines the electronic equilibria of the solid. The electron affinity, on the other hand, is not a thermodynamically controlled quantity.

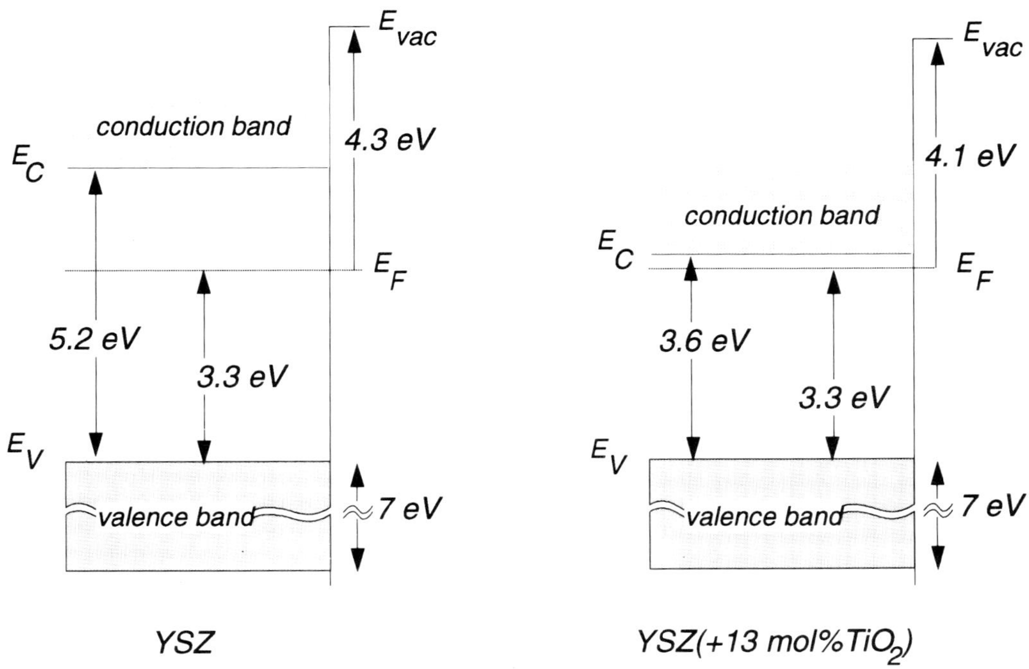

Figure 7 Effect of titania additions on the work function and the band gap of YSZ (values from Ref. 9).

It depends on the polarity of the surface and, hence, on its orientation and on the surface dipole moments of adsorbed gases. The first results for the electron affinity of stabilised zirconia as a function of the yttrium content were obtained in experiments based on thermal emission of electrons.[21] Changes of the work function with oxygen partial pressure were determined with Kelvin's method.[22] Recent results obtained with UPS (UV Photoelectron Spectroscopy) allowed to derive separate values for the work function, electron affinity and position of Fermi level.[5,16]

The use of equilibria with oxygen in the gas phase or with a solid mixed conducting oxide allowed the determination of the Fermi level and the work function in terms of the oxygen activity or partial pressure of oxygen. It was found with Kelvin's technique and UPS that the work function at constant temperature shifts proportionally to $-kT/4 \log p_{O_2}$.[8,22] With UPS, it could be verified that this is due to the partial pressure dependence of the position of the Fermi level as predicted by equation (3).

The experimental results of UPS were used to derive a scale for the oxygen electrode potential in the band scheme of stabilised zirconia (Fig. 8). It gives the position of the Fermi level as a function of the electrode potential. Figure 8 also makes evident why YSZ is a suitable electrolyte material for fuel cells: the partial pressure range from 1 bar to 10^{-25} bar oxygen corresponds to Fermi levels lying in a range of 1 eV around the centre of the band gap, where no localised electronic states (i.e. the band tails) are found.

Stabilised ceria and bismuth oxide are in principle much better oxide ion conductors than zirconia. However, their use is limited by the smaller band gap values which reduce the range of partial pressures of oxygen where electronic conductivity remains small (for instance Bi_2O_3: $E_{gap} = 2.8$ eV). Apart from this, the Fermi level in these solid electrolytes is located nearer to the conduction band edge than in the case of zirconia if compared for the same oxygen partial pressures.[23]

A possible approach could be the attempt to increase the gap by a suitable substitution. In the case of Bi_2O_3, the effect of ZrO_2 and PrO_2 additions has been analysed.[23] The band gap was indeed increased as expected (by about 0.4 eV), but the effect was too small to suppress the electronic conductivity to a larger degree. Similar experiments with ceria have recently led to some success in depressing the electronic conductivity.[24]

5. INTERFACE ELECTROLYTE/ELECTRODE

5.1 Structure and Composition of the Electrode Interface Pt/ZrO_2

The next point concerns the interface between a solid electrolyte and an electronically conducting electrode. First, the system Pt/YSZ is taken as an example illustrating some atomistic properties of the two-phase boundary studied with surface analytical methods. Furthermore, results on the Ag/YSZ interface will be discussed in terms of work function and reaction intermediates of the oxygen electrode reaction.

Figure 2 illustrates the polarisation of a Pt/YSZ interface. Anodic polarisation leads to a downward shift of the Fermi level within the band scheme of the solid electrolyte. For increasing anodic overpotentials, the hole concentration strongly increases near the interface of YSZ and a net oxidation of oxide ions to oxygen atoms occurs, followed by recombination and desorption of oxygen molecules at the three-phase boundary line. Cathodic polarisation, on the other hand, shifts the Fermi level towards the conduction band. For high cathodic overpotentials, electrons are available on localised states in the vicinity of the conduction band edge. Any atomic oxygen is rapidly reduced. Then, oxygen depletion is expected at the interface.

For electrochemical cells with YSZ, electrode potentials are normally applied in a range between 0 V and −1.5 V vs. air. This corresponds to a range of 1.5 eV around the centre of the band gap of YSZ. It corresponds to about 30 orders of magnitude in the partial pressure of oxygen in Fig. 8. Diffuse space charge layers are virtually negligible at the usual temperatures due to the high vacancy concentration of YSZ. Thus, at high overpotentials, quite high gradients of the thermodynamic activity of oxygen must occur at the two-phase boundary. Therefore, the chemical composition, the interface dipole layer and bonding at the interface of electrolyte and electrode must show great changes as a function of polarisation.

In general, the electrolyte/electrode interface is not directly accessible to surface spectroscopic techniques. In view of the small information depth of these techniques,

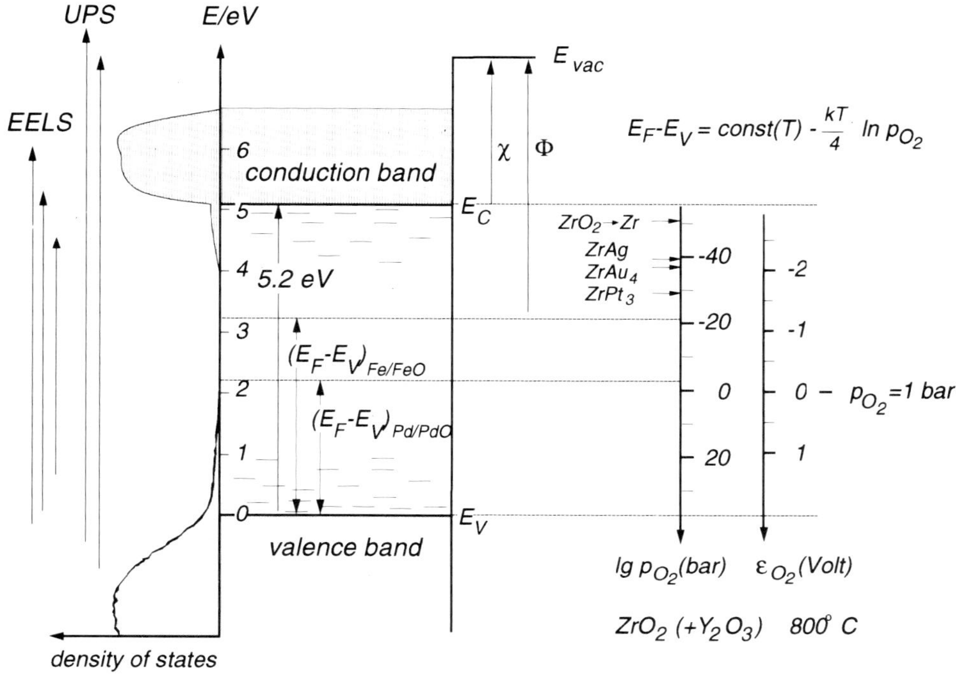

Figure 8 Scales for the electrode potential and the logarithm of oxygen partial pressure in the band diagram of YSZ obtained with UPS and EELS (χ and Φ denote the electron affinity and the work function of electrons at the surface which can also be determined by UPS; detailed results were given in Refs 5, 16).

interface studies have to be carried out on thin films of electrodes on an electrolyte. Figure 9 summarises some of the main results from XPS and ISS as obtained on thin platinum films evaporated on single crystals of YSZ.[8] Reversible changes of the geometry of the electrode film were observed as a function of the electrode potential. At high electrode potentials corresponding to high oxygen partial pressures, small platinum clusters are formed, whereas for low oxygen partial pressures, platinum spreads out covering a larger part of the electrolyte surface. With XPS, at decreasing oxygen partial pressures, i.e. $p_{O_2} < 10^{-10}$ mbar and 1000 K, an increasing intensity of an additional shoulder in the Zr3d peak was found at lower binding energies. This observation indicated a higher electronic charge density at the zirconium ions of the interface as a result of the decreasing surface concentration of oxygen. The reduced zirconium atoms at an oxygen-deficient Pt/YSZ interface lead to a stronger bonding to the platinum metal. This explains the observed better adhesion of platinum on YSZ after a preceding reduction of the YSZ surface. Thus, the primary electrochemical effect of a changing electrode potential on the YSZ/electrode interface is a strong, but reversible, change of the oxygen-to-metal ratio. This is accompanied by changes of the character of bonding and the surface energies.

As these experiments were carried out on relatively thin evaporated platinum layers, the question arises whether these results are valid for sintered paste

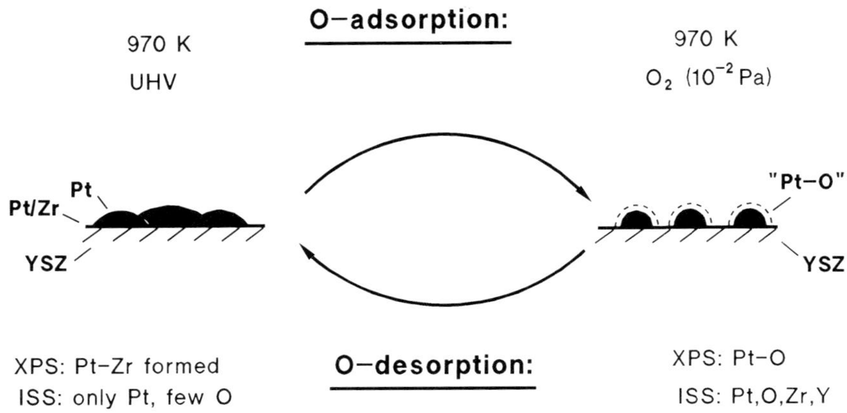

Figure 9 Properties of the Pt/YSZ interface according to results of XPS, ISS as a function of partial pressure of oxygen.

electrodes, too. It is clear that the changes in surface energy have greater consequences for small platinum particles. However, the interface between platinum and YSZ should show the same chemical changes as detected on thin layers. Experiments on Pt paste electrodes often yielded instabilities of the steady-state current–voltage relation which are most likely related to the changes of composition, bonding and adhesion at the Pt/YSZ interface.

5.2 Potential Dependent Changes at Ag/YSZ Electrodes

The last point to be discussed concerns the surface properties of an electrode material in contact with a solid electrolyte and as a function of the electrode potential. Ag/YSZ electrodes are taken as a model system.

Evaporated silver electrodes on YSZ show similar morphological changes after cathodic and anodic polarisation as compared to the Pt/YSZ interface: in open pores of the silver layer, insulated small silver clusters and larger spherical silver particles appear after cathodic and anodic polarisation (for instance after anodic polarisation with 900 mV for 10 minutes at 500°C).[10] They are dispersed within the pores over the entire free zirconia surface. No silver particles were detected in the pores after merely heating up the galvanic cell without polarisation. The appearance of the silver particles differed between anodic and cathodic experiments. A larger amount of very small silver particles and a more uniform distribution over the pores was found for anodic polarisation. Less but larger particles were detected after cathodic polarisation.

It is likely that these observations can also be explained by potential-dependent changes of the bonding and composition of the Ag/YSZ interface. Contrary to the platinum, silver metal is able to dissolve atomic oxygen in the bulk. High anodic potentials may also cause the formation of silver oxide at the Ag/YSZ interface and on the Ag surface. The latter has been detected in recent XPS experiments on anodically polarised silver electrodes.[10] Silver atoms, in addition, seem to be relatively mobile on the YSZ surface.

A net cathodic or anodic polarisation, for instance, leads to a difference of the Fermi levels of the Ag electrode and the YSZ electrolyte which directly corresponds to the overvoltage at the silver electrode. If insulated silver particles appear, they must be currentless and, therefore, their Fermi level should be equal to or at least nearly equal to that of the solid electrolyte. Any overvoltage of the coherent silver layer, then, must lead to a corresponding difference between the Fermi levels of the silver layer and the insulated silver particles. This was indeed the case as shown by UPS results.[10]

5.3 UPS and Work Function of Silver Electrodes on YSZ

The work function of an electrode is a suitable indicator for changes of composition and for adsorbed species that occur at the electrode surface after a change of electrode potential. As demonstrated by Vayenas and his coworkers, the work function (and therefore its change with the electrode potential) is also strongly related to the catalytic activity of an electrode surface.[25] For metal electrodes, the primary cause for a changing work function must be attributed to the influence of adsorbed gas molecules on the surface dipole moment. The work function of a mixed conducting electrode, however, depends in addition on the stoichiometric composition which can influence the electron affinity as well as the position of the Fermi level. Such behaviour was found for oxides like $La_{1-x}Sr_xCoO_3$ on YSZ,[26] but also for metals like silver or palladium which can dissolve oxygen or hydrogen.

Figure 10 shows UPS results obtained with evaporated silver electrodes on YSZ under cathodic and anodic polarisation. A detailed discussion of the spectra can be found in Ref 10. The intensity between 0 (= Fermi level E_F^o) and about 10 eV is due to electrons on occupied states in the conduction band of silver. Atomically adsorbed oxygen mainly leads to an additional intensity near 3.4 eV in Fig. 9 for anodic polarisation. Under cathodic conditions, a slight increase of intensity was observed at binding energies between 8.5 and 9.5 eV, which we attributed to hydroxyl species formed on the silver surface under cathodic conditions. This is in agreement with UPS results for hydrogen-treated silver foils, where the presence of surface hydroxyl species led to the same signal [see Ref. 27 and references therein].

The experimentally obtained total width $E_F^o - E^a$ of the UPS spectrum of silver contains the work function of electrons Φ according to

$$E_F^o - E^a = h\nu - \Phi = 21.21\ \text{eV} - \Phi$$

E^a corresponds to the main cut-off edge corresponding to secondary electrons of the polarised silver electrode with the lowest kinetic energies (see Fig. 10a). Note, however, that a small shoulder corresponding to a second cut-off edge at higher binding energies is found in the cathodic UPS spectra (denoted by E^b in Fig. 10a). Because no current flows through the silver particles, the spectrum of the silver

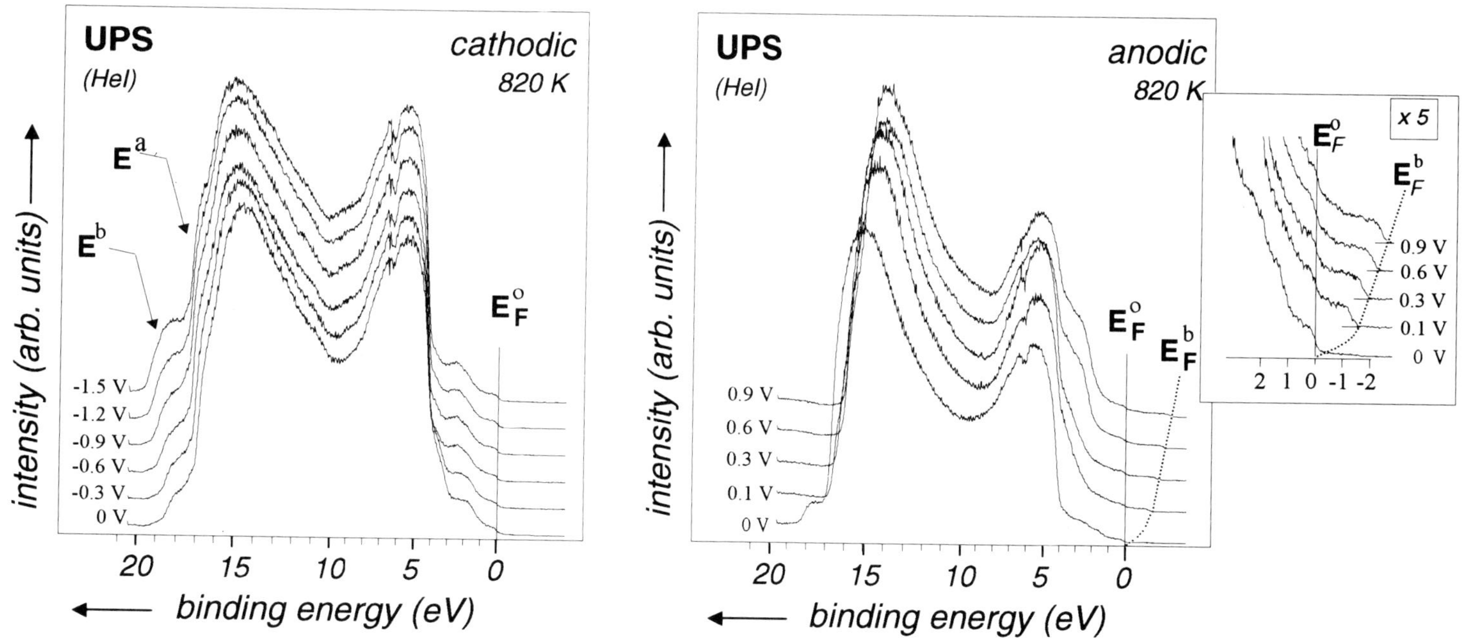

Figure 10 Spectra obtained with UPS for evaporated silver electrodes on YSZ as a function of electrode potential (E_F^o – Fermi energy of the bulk silver electrode, E_F^b – Fermi energy of the insulated silver particles within pores of the bulk electrode material, E^a – secondary electron cut-off for the bulk electrode material, E^b – cut-off for the insulated silver particles): the voltages correspond to the applied cell voltage of the silver electrode vs. a Pd/PdO reference at the back contact of YSZ.

particles is shifted along the energy scale with respect to the main spectrum of the bulk silver electrode by the energy difference $\Delta E = -e \cdot U$ (with U = overvoltage). For cathodic polarisation, the Fermi level of the silver particles and their UP spectrum are shifted to higher binding energies than that of the polarised silver electrode causing the second cut-off edge in Fig. 10. For anodic conditions, on the other hand, the inverse shift leads to the appearance of a second Fermi edge (E_F^b in Fig. 10b) above E_F^o which marks the constant Fermi level of the silver electrode.

Figure 11 shows the results for the work function Φ of silver electrodes as a function of the applied cell voltage U. The work function linearly changes with the overvoltage at the silver electrode. It increases from about 3.8 eV under high cathodic polarisation to about 5.4 eV under high anodic polarisation. The relative change of Φ with the cell voltage U is smaller for cathodic polarisation and largest at low anodic polarisation. The beginning anodic polarisation is characterised by the appearance of a larger amount of adsorbed oxygen atoms and the disappearance of hydroxyl groups. Adsorption of OH should decrease the work function, because its dipole moment is directed towards the silver surface and thus lowers the electron affinity. A partial negative charge of the adsorbed oxygen atoms together with a corresponding positive image charge at the silver surface must lead to an oppositely oriented surface dipole. Thus, an increasing concentration of adsorbed oxygen atoms and a decreasing one of OH have the same effect on the work function. The observed changes of the work function with the electrode potential were completely reversible.

5.4 Adsorption and Composition of Electrode Surfaces

The mechanism of oxygen reduction or formation of oxygen at electrodes on stabilised zirconia has been the subject of many investigations. XPS allows us to study oxygen species in different bonding states at surfaces. Some recent results are discussed in the following for the oxygen electrode reaction at silver on YSZ. A thorough discussion can be found in Ref. 10.

Oxygen is adsorbed dissociatively at the silver surface, dissolves as interstitial oxygen in the bulk of the silver electrode and diffuses through the bulk towards the zirconia/silver interface where it is reduced to oxide ions. Applying an anodic polarisation will lead to a steady-state concentration gradient of oxygen through the silver layer. At the surface of silver, a stationary oxygen coverage will develop that is determined by the balance between surface desorption as O_2 and bulk diffusion of O atoms to the surface. Thus, the surface coverage by oxygen which was studied by UPS and XPS is far from equilibrium conditions. Nevertheless, a large relative change of oxygen coverage can be produced at the surface by varying the electrode potential of silver from about -1.5 V to $+1.5$ V versus a reference electrode.

By applying a potential to the silver electrode, we found reversible changes of the XP-spectra in the O1s region, whereas the Ag3d emission showed no changes in binding energy or shape. XPS difference spectra of the O1s signal for anodic as well as for cathodic polarisation are shown in Fig. 12. To obtain the difference spectra, first the contribution due to O1s(ZrO_2), whose intensity and binding energy were

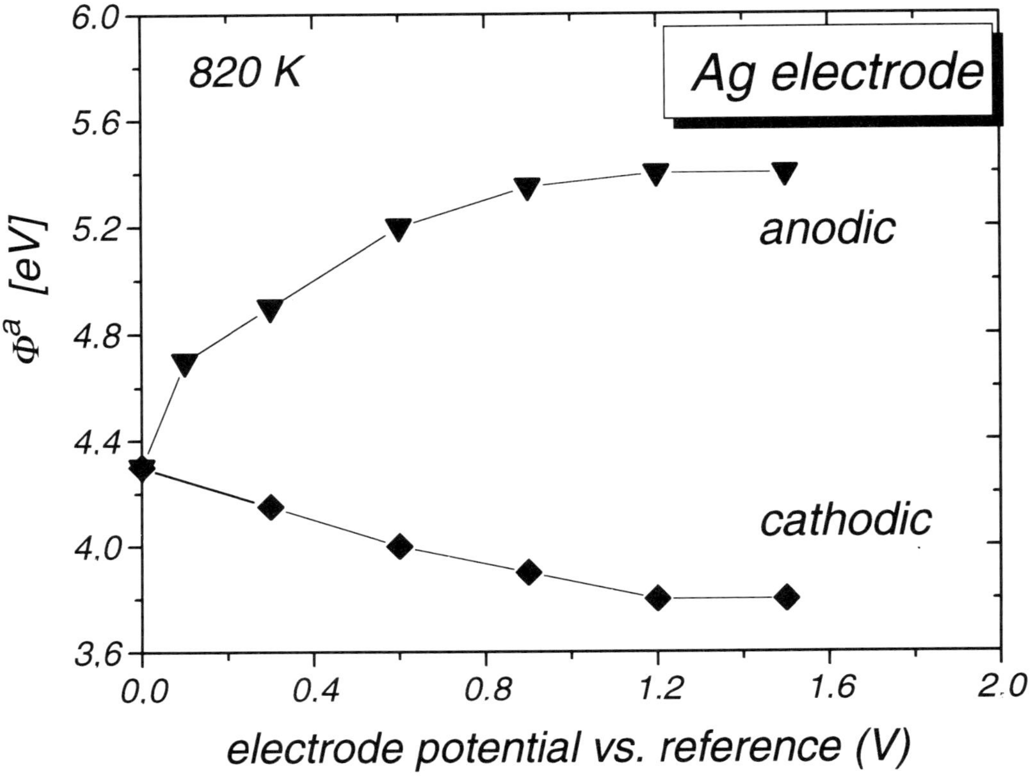

Figure 11 Work function Φ^a of silver electrodes (a denotes bulk material) on YSZ as a function of electrode potential (only absolute values for cathodic and anodic polarisation).

derived from the intensity and position of the measured Zr3d peak in each spectrum, was subtracted. After this correction, the difference spectra in Fig. 12 are obtained by subtracting from each O1s spectrum for $U \neq 0$ V the O1s signal for the short-circuited cell at $U = 0$ V.

An anodic polarisation gives the expected increase in intensity of the O1s signal due to the higher oxygen coverage at the silver surface. However, a considerable O1s intensity is still observed for cathodic polarisation of -1 V. One can conclude from the width of the O1s signal (about 4–5 eV, see Fig. 12) that at least three or more oxygen species are involved. A clean surface of a YSZ single crystal, for instance, gives an O1s signal with a full width at half maximum of only 2 eV.

The O1s signal intensity increases at low binding energies around 529.6 eV with increasing anodic polarisation, whereas the intensity at higher binding energies around 532.5 eV slightly decreases. A similar change of the O1s intensity in opposite directions at high and low binding energies is found for cathodic polarisation. Increasing cathodic polarisation leads to a further decrease of the signal near 529.6 eV and an increase near 532.5 eV.

The increasing intensity at low binding energy for anodic polarisation must be

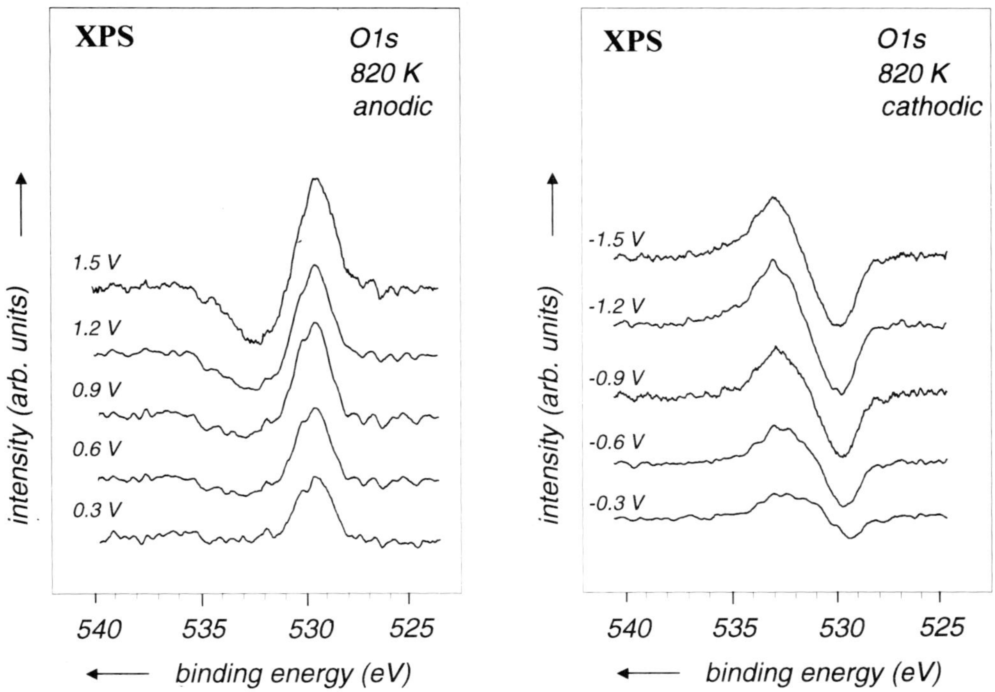

Figure 12 XPS difference spectra of the O1s signal for the galvanic cell Pd,PdO/YSZ/Ag under anodic and cathodic polarisation at 820 K. In each case, the intensity of the zero voltage spectrum has been subtracted. The spectra were also corrected for the contribution of oxide ions of the free parts of the YSZ surface.

attributed to atomic oxygen as an intermediate species which is continuously supplied by the electrode reaction, that is, by the transport of oxygen through the silver electrode to the surface. The increasing part of the O1s intensity centred at 529.6 eV is likely to contain at least two different oxygen species.

The inverse changes in the O1s signal intensity around 532.5 eV are more difficult to explain. The most important point is that the intensity increases for reducing conditions and decreases under oxidising conditions. This excludes an explanation based on oxidised species like O_2, O_2^- which are known to produce O1s signals in that range of binding energies. A signal component due to molecular, adsorbed O_2 or O_2^- at binding energies higher than 532 eV should solely appear for anodic polarisation, for which it was not found.

Hydroxyl groups are assumed to be the species containing reduced oxygen that could explain the signal for cathodic polarisation. The residual gas in the UHV as well as the materials of the galvanic cell always contain water traces. Water will be reduced at the interface silver/solid electrolyte under cathodic polarisation. The resulting adsorbed hydrogen radicals are very reactive and should react with any traces of oxygen to give hydroxyl groups. A certain solubility of hydrogen in silver may be involved, too, although the formation of OH on the silver surface might also be explained by surface diffusion of adsorbed hydrogen formed at the three-phase line vacuum-silver-solid electrolyte. The O1s binding energy of OH_{ad} on silver is usually found near 532 ± 1 eV (see e.g. Ref. 30). To conclude, we assign the signal at 532.5 eV to hydroxyl species. This explains the reversible increase in O1s intensity at 532.5 eV from anodic to cathodic polarisation. Extensive *in-situ* XPS studies of anodic oxygen formation at metal/YSZ electrodes have been made up to now only on the electrode systems Ag/YSZ[6,10] and Pt/YSZ.[31]

6. OUTLOOK

Our results demonstrate that important properties of surfaces and interfaces of solid electrolyte and electrode surfaces can be studied *in-situ* with surface spectroscopic techniques. They make possible a more detailed knowledge of the atomistic structure and composition of surfaces and interfaces, as well as of the relevant electrode processes. This is desirable in order to improve the selectivity and kinetics of electrode processes in sensors and other electrochemical devices. The analysis of intermediate reactive oxygen species by XPS in principle gives insight into the mechanism of the electrode reaction.

In-situ photoelectron spectroscopy will be a very interesting technique if applied to oxidic electrode materials. Some results were recently obtained for oxide electrodes like $La_{1-x}Sr_x(Co,Ni)O_3$[7] and $La_{1-x}Sr_xMnO_3$.[28] As in the case of metal electrodes, the work function of oxidic mixed conducting electrodes changes with the applied electrode potential. Broad XPS O1s signals on these materials show that under polarisation different oxygen species are present at the surface. Surface and bulk hydroxyl groups seem to play a role in all these oxides. Hydroxyl groups and water desorption are still found on these surfaces at temperatures as high as 800°C. Reduction or oxidation of the electrode material itself makes the conditions more complicated than in the case of noble metals. Recent experiments on $La_{1-x}Sr_xCoO_3$, for instance, gave evidence for segregation of $SrCO_3$ and $Sr(OH)_2$.[29]

7. REFERENCES

1. W. F. Egelhoff jr.: *Surface Science Reports*, 1987, **6**, 253–415.
2. G. Ertl and J. Küppers: *Low Energy Electrons and Surface Chemistry*, 2nd ed., VCH Weinheim 1985.
3. M. Levy, J. Fouletier and M. Kleitz: *J. Electrochem. Soc.*, 1988, **135**, 1584–1589.
4. M. Kleitz, E. Siebert, P. Fabry and J. Fouletier: *Sensors – A Comprehensive Survey*, W. Göpel, J. Hesse, J. N. Zemel (eds), VCH Weinheim 1991, 341–428.
5. H.-D. Wiemhöfer, S. Harke and U. Vohrer: *Solid State Ionics*, 1990, **40/41**, 433–439.
6. T. Arakawa, A. Saito and J. Shiokawa: *Applications of Surface Science*, 1983, **16**, 365.
7. H.-D. Wiemhöfer: *Ber. Bunsenges. Phys. Chem.*, 1993, **97**, 461–469.
8. K. Schindler, D. Schmeisser, U. Vohrer, H.-D. Wiemhöfer and W. Göpel: *Sensors and Actuators*, 1989, **17**, 555.
9. U. Vohrer, H.-D. Wiemhöfer, W. Göpel, B. A. van Hassel and A. J. Burggraaf: *Solid State Ionics*, 1993, **59**, 141–149.

10. W. Zipprich, H.-D. Wiemhöfer, U. Vohrer and W. Göpel: *Ber. Bunsenges. Physikal. Chem.*, 1995, **99**, 1406–1413.
11. H.-D. Wiemhöfer, U. Vohrer and W. Göpel: *Mat. Sci. Forum*, 1991, **76**, 265–268.
12. U. Vohrer, H.-D. Wiemhöfer, W. Göpel, F. Schilling and J. Arndt: *Sensors Actuators*, 1991, **B4**, 411–416.
13. A. J. A. Winnubst, P. J. M. Kroot and A. J. Burggraaf: *J. Phys. Chem. Solids*, 1983, **44**, 955–960.
14. B. C. H. Steele and E. P. Butler: *Brit. Ceram. Proc.*, 1985, **36**, 45–55.
15. A. J. Burggraaf, M. Van Hemert, D. Scholten and A. J. A. Winnubst: *Reactivity of Solids*, P. Barret and L.-C. Dufour (eds), Elsevier, Amsterdam, 1985, pp. 797–802.
16. H.-D. Wiemhöfer and U. Vohrer: *Ber. Bunsenges. Phys. Chem.*, 1992, **96**, 1646–1652.
17. J.-H. Park and R. N. Blumenthal: *J. Electrochem. Soc.*, 1989, **136**, 2867–2876.
18. P. W. Anderson: *Phys. Rev.*, 1958, **B109**, 1492–1505.
19. N. Mott, M. Pepper, S. Politt, R. H. Wallis and C. J. Adkins: *Proc. R. Soc. London*, 1975, **A345**, 169–205.
20. S. S. Liou and W. L. Worrell: *Appl. Phys.*, 1989, **A49**, 25–31.
21. P. Odier and J. P. Loup: *Proc. 9th Int. Symp. Reactivity of Solids*, K. Dyrek, J. Haber and J. Nowotny (eds), Elsevier, Amsterdam, 1982, pp. 458–473.
22. J. Nowotny and M. Sloma: *J. Phys. (Paris) Colloq.*, 1986, **47**, C1-807–811.
23. P. Shuk, H.-D. Wiemhöfer and W. Göpel: in preparation.
24. H. L. Tuller: this conference.
25. S. Ladas, S. Bebelis and C. G. Vayenas: *Surface Science*, 1991, **251/252**, 1062–1068.
26. W. Zipprich and H.-D. Wiemhöfer: unpublished results.
27. C. Rehren, M. Muhler, X. Bao, R. Schlögl and G. Ertl: *Z. Phys. Chem.*, 1991, **174**, 11–52.
28. H. Y. Lee, W. S. Cho, S. M. Oh, H.-D. Wiemhöfer and W. Göpel: *J. Electrochem. Soc.*, 1995, **142**, 2659–2664.
29. G. Reinhardt, H.-D. Wiemhöfer and W. Göpel: in preparation.
30. M. Barteau and R. Madix: *Surf. Sci.*, 1990, **140**, 108–122.
31. S. Ladas, S. Kennou, S. Bebelis and C. G. Vayenas: *J. Phys. Chem.*, 1993, **97**, 8845–8848.

Oxygen Ion Conductivity in Brownmillerite-structured Oxides, $Ba_2In_{2-x}M_xO_5$, where M = Y, Yb, Sc, Ga

C. A. J. FISHER, B. DERBY and R. J. BROOK
Department of Materials, University of Oxford, Parks Road, Oxford, OX1 3PH, U.K.

ABSTRACT
Brownmillerite-structured oxides ($A_2B_2O_5$) are promising fast oxygen ion conductors because of the large number of oxygen vacancies inherent in the structure. However, at temperatures of practical interest (<800°C) the vacancies are ordered and do not take part in rapid ionic conductivity. $Ba_2In_2O_5$ is a brownmillerite oxide with an order–disorder transition temperature, $T_t = 880$°C and a conductivity, $\sigma \approx 0.1$ S.cm^{-1} at 1000°C. The substitution of small amounts of isovalent elements on the In site, $Ba_2In_{2-x}M_xO_5$, where M = Y, Yb, Sc, Ga and $x = 0.1, 0.2$, has shown that the transition temperature can be increased or decreased relative to the undoped material. Trivalent ions larger than In^{3+} are found to increase the transition temperature, whereas substitution by the smaller Sc^{3+} and Ga^{3+} ions lowers T_t. The change in transition temperature is correlated with changes in the unit cell free volume.

1. INTRODUCTION

Ceramics with a high proportion of oxygen vacancies are expected to display enhanced oxygen ion conductivity. The brownmillerite structure is essentially a superstructural form of the cubic perovskite structure with one sixth of the oxygen atoms removed (Fig. 1), and thus contains a large number of intrinsic oxygen vacancies.[1] In the brownmillerite structure, however, the vacancies order along the [101] direction in alternate layers, producing an orthorhombic unit cell. This ordering effectively traps the vacancies and stops them from contributing to any rapid ionic conduction.[2]

Above a transition temperature, T_t (typically 800–1000°C), the brownmillerite structure reverts to a highly oxygen-deficient perovskite with disordered vacancies.[3] For $Ba_2In_2O_5$, T_t has been reported to be 880°C in air[4] and 930°C at low oxygen partial pressures.[5] Above T_t, fast oxygen ion transport takes place.

Several strategies have been employed for lowering the order–disorder transition temperature to allow fast ion conduction at lower temperatures.[5–7] Goodenough and co-workers originally suppressed oxygen vacancy ordering by introducing an extra octahedral layer into the structure through substitution of quadrivalent ions for In^{3+}.[5] These intergrowth structures, such as $Ba_3In_2ZrO_8$, although successfully removing the vacancy ordering, suffer from degradation due to water absorption below 400°C.[8]

Substitution of Ba^{2+} with Pb^{2+} by Kuramochi and co-workers removed the transition temperature entirely in the composition $(Ba_{0.80}Pb_{0.09})_2In_2O_5$.[7] They

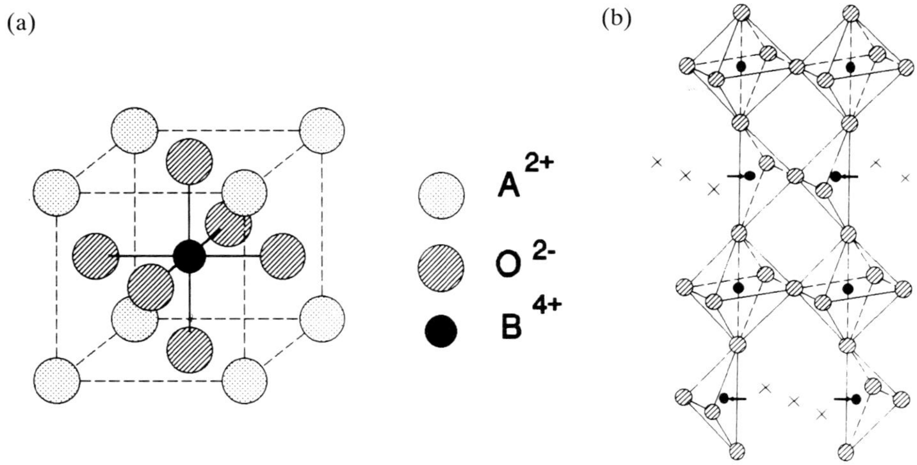

Figure 1 (a) Ideal perovskite structure, ABO_3; (b) ideal brownmillerite structure, $A_2B_2O_5$ (× = oxygen vacancy, A cations removed for clarity).

found that the unit cell volume increased with increasing lead content, while the transition temperature decreased. Substitution with Sr^{2+}, roughly the same as size Pb^{2+}, had the same effect on the transition temperature, but with lower conductivities, suggesting that the polarisability of Pb^{2+} plays a role in promoting anion transport. However, because of lead's multiple valencies, electronic contributions to the total conductivity are likely to be significant under variable redox conditions.

Increased conductivities in barium indate ceramics by doping on the B (In) site have also been reported.[6] The purpose of this work is to examine the role of ionic size on T_t and to show that the order–disorder transition temperature can be controlled by manipulation of the trivalent cation (B) sublattice.

2. EXPERIMENTAL

2.1 Sample Preparation

Precursor powders for samples of $Ba_2In_{2-x}M_xO_5$, M = Y, Yb, Sc or Ga, and $x = 0$, 0.1, 0.2, were prepared by a combustion reaction route. Nitrates of each metal were dissolved in water and urea (CON_2H_4) added to give a reducing elements/oxidising elements ratio of 1.20. The solution was heated to ≈300°C on a hot-plate to evaporate the water solvent. When almost dry, the remaining mixture ignited, burning with a bright orange flame and producing fine, aerated powder. These powders were calcined at 900°C for 2 h to ensure complete reaction of the mixture.

Calcined powders were ball-milled for 2 h using ZrO_2 media and ethanol and subsequently dried and passed through a 75 μm sieve. Bars and pellets of the appropriate composition could then be produced with high green density by pressing at 210 MPa. Samples were sintered at 1350°C for 15 h in air. Final density and open porosity were determined by pycnometric measurements in which *n*-octanol was used as the liquid medium ($\rho = 0.827\ \mathrm{g/cm^3}$) and air evacuated using a vacuum pump.

2.2 Characterisation

Phase identification was performed by X-ray diffraction (XRD) using Cu Kα radiation from a Philips diffractometer. XRD patterns were indexed and unit cell lattice parameters calculated using the DRAGON and CELL programs.

Four-point probe DC conductivity measurements were performed on sintered bars. Platinum paste (Engelhard) was used to attach platinum wire electrodes along the length of each bar. Samples were heated over the temperature range 600 –1100°C and resistance measurements taken using a Keithley 196 System Digital Multimeter. Measurements were also taken on cooling from 1100 to 600°C for several samples to determine if any hysteresis was taking place. The transition temperatures were measured as the mid-point between the $1/T_{\mathrm{onset}}$ and $1/T_{\mathrm{offset}}$. The activation energy for fast ion conduction, E_a, was calculated from the slope of the conductivity curve above the T_t.

AC impedance measurements of pellets were taken over a temperature range of 600–1000°C in air using a Schlumberger SI 1260 impedance analyser. A frequency range of 10^7–1 Hz was used. $Ba_2In_2O_5$ was also investigated under Ar($pO_2 \approx 10^{-6}$) to examine the effect of lower oxygen partial pressures.

3. RESULTS AND DISCUSSION

3.1 Phase Identification

The X-ray diffraction pattern of single phase $Ba_2In_2O_5$ is shown in Fig. 2. It could be indexed according to an orthorhombic cell and the lattice parameters were calculated to be $a = 6.08$, $b = 16.80$, $c = 5.89$, using Si as an internal standard. These are in excellent agreement with those reported by Goodenough *et al.*,[4] $a = 6.08$, $b = 16.79$, $c = 5.89$ Å, with a slight extension of the c parameter.

XRD results for the doped samples showed that they were essentially single phase, although small amounts of the dopant oxides were detectable. The densities of all samples except those doped with Ga were calculated to be ≈90% of theoretical. The Ga doped samples sintered poorly and dense samples could not be produced. It is likely that Ga^{3+} (ionic radius 0.62 Å compared to 0.80 Å for In^{3+})[9] is too small to be accommodated on the $Ba_2In_2O_5$ lattice in the amounts substituted. Consequently, sufficiently strong bars of Ga doped compositions could not be produced, and conductivity measurements on these materials were performed by impedance spectroscopy only.

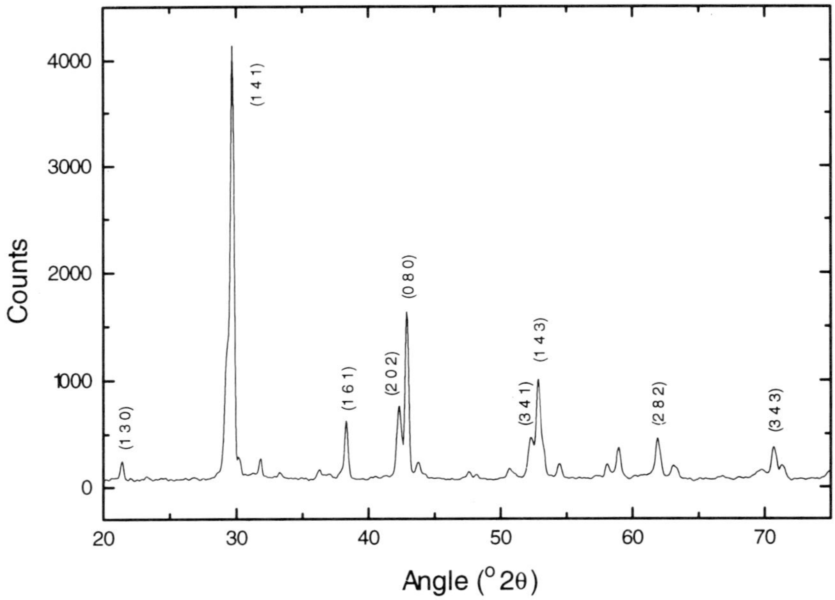

Figure 2 XRD pattern of $Ba_2In_2O_5$ at 25°C.

3.2 Electrical Conductivity

The transition temperature of $Ba_2In_2O_5$ was found to be 880°C in air by both DC and AC techniques. Under Ar, however, the transition temperature increased to 925°C and the conductivity was noticeably less below T_t. This indicates that atmospheric oxygen plays a role in the disordering process.

DC conductivity measurements on samples doped with Y, Yb and Sc are shown in Figs 3, 4 and 5 respectively. The following features can be discerned:

- Doping with ions larger than In^{3+} increases the transition temperature. The curves for Y and Yb are very similar, which is to be expected because of the similarity in their ionic sizes, 0.89 and 0.87 Å respectively,[9] and both raise T_t by ≈80°C.
- The total conductivity of Y and Yb doped samples above T_t was increased by a factor of 2 relative to the undoped material, but suppressed below T_t. Increasing the dopant level appears to have little effect above the transition temperature, but, paradoxically, the samples with higher dopant level displayed the higher conductivity below T_t.
- Doping with the smaller Sc^{3+} ion (ionic radius 0.85 Å)[9] decreased the transition temperature. T_t was decreased to 825°C for $x = 0.1$ and 770°C for $x = 0.2$. A distinctive feature of the $x = 0.2$ curve is the increased width of the transition region compared to the other samples. In other words, disordering of the oxygen sublattice is beginning at much lower temperatures.
- The gradients of the curves above T_t for Sc doped samples are almost identical to

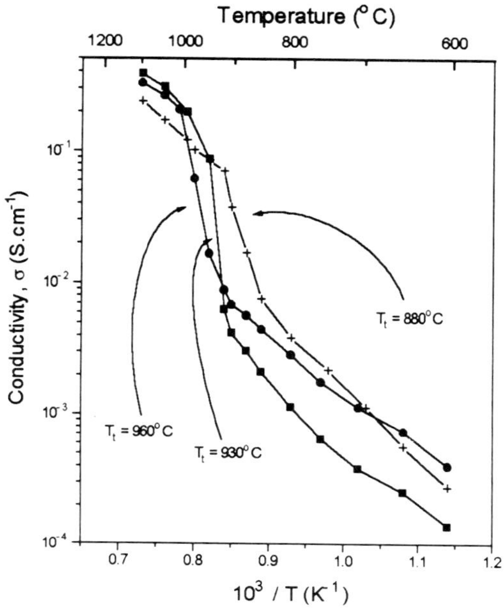

Figure 3 $Ba_2In_{2-x}Y_xO_5$ (+ $x = 0$, ■ $x = 0.1$, ● $x = 0.2$).

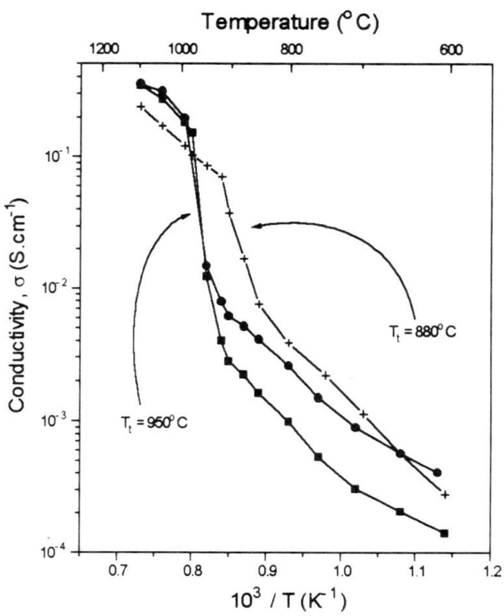

Figure 4 $Ba_2In_{2-x}Yb_xO_5$ (+ $x = 0$, ■ $x = 0.1$, ● $x = 0.2$).

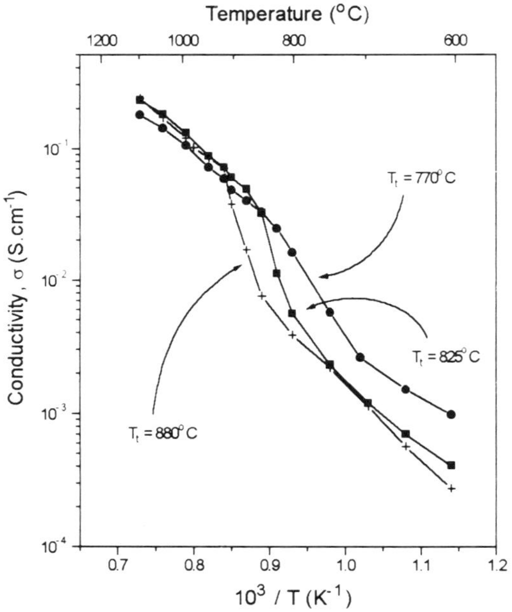

Figure 5 $Ba_2In_{2-x}Sc_xO_5$ (+ $x = 0$, ■ $x = 0.1$, ● $x = 0.2$).

that of the undoped sample, indicating that the dopant has no effect on the activation energy for fast ionic conduction at these temperatures. E_a was calculated to be ≈ 1.0 eV, which is comparable to those reported for fluorite structured oxygen ion conductors.[1]

A comparison between DC conductivity measurements made on heating and cooling is shown in Fig. 6. Although very little difference is observed above T_t, below T_t the conductivity is much less when cooling from elevated temperatures. If the sample is reheated again, the low temperature conductivity does not reach its former levels, although the transition occurs at the same temperature and the high temperature conductivity is the same. This ageing phenomenon in the low temperature brownmillerite structure was most noticeable in the Ga doped samples investigated by AC impedance spectroscopy. The transition temperatures of $Ba_2In_{1.9}Ga_{0.1}O_5$ and $Ba_2In_{1.8}Ga_{0.2}O_5$ were found to be 890°C and 850°C respectively. However, the behaviour of the Ga doped samples was complicated by the presence of impurities and their poor density. The conductivity curves were not always reproducible and hence these values of T_t are associated with large errors.

3.3 *Unit Cell Free Volume*

The unit cell free volume is defined here as the difference between the total unit cell volume and the sum of volumes occupied by individual ions in the unit cell. It is also

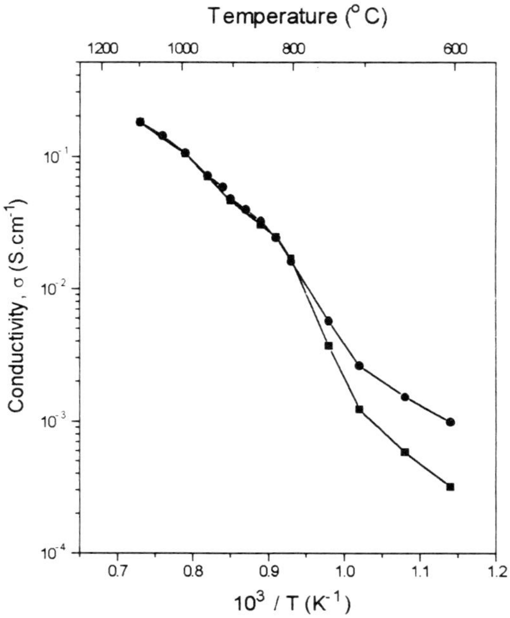

Figure 6 Ageing of $Ba_2In_{1.8}Sc_{0.2}O_5$ (● heating, ■ cooling).

referred to as the lattice free volume, and has been correlated with activation energies in perovskite oxides.[10] Using lattice parameters calculated from the XRD patterns, and ionic radii from Shannon,[9] values for the unit cell free volume were calculated for each composition.

A plot of transition temperature versus unit cell free volume is shown in Fig. 7. An overall trend is apparent. The larger the unit cell free volume, the lower the transition temperature. Similarly, the smaller the free volume, the higher the transition temperature. The Ga doped samples appear as anomalies, but, as stated previously, reliable values of T_t for these materials were difficult to obtain. Nevertheless, the free volumes calculated for the Ga doped samples are sufficiently close to that of $Ba_2In_2O_5$ to expect their transition temperatures to be near 880°C.

A similar result was found by Kuramochi *et al.* when they observed that the transition temperature decreases with increasing unit cell volume in Pb and Sr doped $Ba_2In_2O_5$ ceramics.[7] Sr^{2+} and Pb^{2+} are both smaller than Ba^{2+}, but result in larger unit cell volumes. These factors combine to give a larger unit cell free volume compared with the undoped material. Large values of unit cell free volume are associated with higher oxygen ion conductivities because the oxygen ions can pass more easily from site to site in the lattice[10] (this is facilitated further by the high polarisability of Pb^{2+} in Pb doped samples). Larger unit cell free volumes will decrease the order–disorder transition temperature because movement of oxygen ions disorders the oxygen sublattice. When a critical number of anions has moved

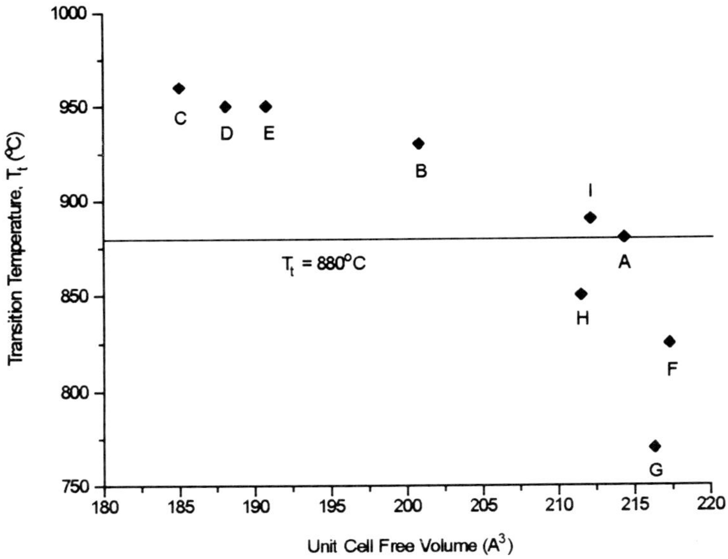

Figure 7 Transition temperature versus unit cell free volume: A: $Ba_2In_2O_5$. B: $Ba_2In_{1.9}Y_{0.1}O_5$. C: $Ba_2In_{1.8}Y_{0.2}O_5$. D: $Ba_2In_{1.9}Yb_{0.1}O_5$. E: $Ba_2In_{1.8}Yb_{0.2}O_5$. F: $Ba_2In_{1.9}Sc_{0.1}O_5$. G: $Ba_2In_{1.8}Sc_{0.2}O_5$. H: $Ba_2In_{1.9}Ga_{0.1}O_5$. I: $Ba_2In_{1.8}Ga_{0.2}O_5$.

to the higher energy ordered vacancy sites,[3] transformation from the ordered brownmillerite structure to a disordered perovskite structure will occur.

4. CONCLUSION

The brownmillerite oxides are promising fast ion conductors with rapid oxygen transport above the order–disorder transition temperature. $Ba_2In_2O_5$ has been synthesised with a transition temperature of 880°C and conductivity of 0.1 S/cm at 1000°C. Substitution of the trivalent cation with isovalent ions of larger size, although increasing the net conductivity at elevated temperatures, increases the transition temperature. Substitution with the smaller Sc^{3+} ion, however, can lower the transition temperature by >100°C to 770°C. These changes are correlated to changes in the unit cell free volume of the brownmillerite phases. An increase in the unit cell free volume is associated with a lower T_t, since less energy is required for oxygen ion transport and disordering of the oxygen sublattice. Conversely, a smaller free volume leads to a higher T_t.

ACKNOWLEDGEMENTS

The authors would like to thank the Royal Society of London for sponsoring C.F. on a Rutherford Scholarship while doing this work.

REFERENCES

1. B. C. H. Steele: *Mater. Sci. & Eng.*, 1992, **B13**, 79–87.
2. S. Adler, S. Russek, J. Reimer, M. Fendorf, A. Stacy, Q. Huang, A. Santoro, J. Lynn, J. Baltisberger and U. Werner: *Sol. State Ionics*, 1994, **68**, 193–211.
3. S. B. Adler, J. A. Reimer, J. Baltisberger and U. Werner: *J. Am. Chem. Soc.*, 1994, **116**, 675–681.
4. T. Mitamura, H. Ogino and H. Kobayashi: *J. Am. Cer. Soc.*, 1993, **76**(8), 2127–2128.
5. J. B. Goodenough, J. E. Ruiz-Diaz, and Y. S. Zhen: *Sol. State Ionics*, 1990, **44**, 21–31.
6. M. Schwartz, B. F. Link and A. F. Sammells: *J. Electrochem. Soc.*, 1993, **140**(4), L62–L63.
7. H. Kuramochi, T. Mori, H. Yamamura, H. Kobayashi and T. Mitamura: *J. Cer. Soc. Japan*, 1994, **102**(12), 1159–1162.
8. J. B. Goodenough, A. Manthiram, P. Paranthaman and Y. S. Zhen: *Sol. State Ionics*, 1992, **52**, 105–109.
9. R. D. Shannon: *Acta Crystallogr.*, 1976, **A32**, 751–767.
10. A. F. Sammells, R. L. Cook, J. H. White, J. J. Osborne and R. C. MacDuff: *Sol. State Ionics*, 1992, **52**, 111–123.

Influence of Grain Size on the Oxygen Ion Conductivity in $Zr_{0.85}Y_{0.15}O_{1.93}$

R. VASSEN, D. STÖVER[1]
L. G. J. DE HAART and M. CAPPADONIA[2]
[1]*Institut für Werkstoffe der Energietechnik*
[2]*Institut für Energieverfahrenstechnik, Forschungszentrum Jülich GmbH, D-52425 Jülich, Germany*

ABSTRACT
$Zr_{0.85}Y_{0.15}O_{1.93}$ is a promising oxygen ion electrolyte material for solid oxide fuel cells (SOFC). But a still further improvement of the ion conductivity of this material would be favourable in order to reduce the operation temperature. A possible way might be the reduction of grain size down to the nanometer size range. In this grain size regime oxygen ion transport along grain boundaries will play an increasing role and as a result an increase in ion conductivity compared to coarse grained zirconia might be possible.

Fine grained and nanosized zirconia in the grain size regime below 500 nm has been produced by pressure-assisted sintering techniques. Hot pressing at pressures of 110 MPa led to dense samples at 1100°C with a final grain size of about 110 nm. For oxygen ion conductivity measurements (four-point electrical conductivity measurements and impedance spectroscopy) disk-shaped specimens have been produced with different grain sizes. For grain sizes below 200 nm a decrease in the activation energy of the ionic conductivity was observed.

1. INTRODUCTION

Nanocrystalline materials offer a wide range of new and promising properties.[1] Especially the high number of diffusion paths within the grain boundaries are likely to offer interesting applications of these materials for processes in which materials transport is of essential importance as in the electrolyte of the SOFC.

Several efforts have been made to produce nanocrystalline zirconia,[2,3,4] indicating that the production of dense cubic zirconia samples with grain sizes below 200 nm is much more difficult than the production of ZrO_2/3 mol% Y_2O_3 samples in this size range. As a result the ion conductivity measurements performed at fine grained zirconia samples are mainly made at ZrO_2/3 mol% Y_2O_3 samples (Chen,[5] 100–130 nm; Mayo,[6] 80–100 nm) or at relatively coarse grained cubic zirconia samples (Verkerk *et al.*,[7] 360 nm). The present investigation is concentrated on the production of fine grained cubic zirconia and the investigation of the ionic conductivity.

2. EXPERIMENTAL

Samples were prepared from a ZrO_2/8 mol% Y_2O_3 powder produced by Tosoh Corporation (see Table 1). A few samples were made from ultrafine zirconia powder

Table 1
Characterisation of the powders used

Powders	Producer	Particle size (nm)	Grain size (nm)	Preparation technique
ZrO_2/8 mol% Y_2O_3	Tosoh Corporation, Japan	300	25	Sol gel
ZrO_2/3 mol% Y_2O_3	Nanophase Technology, Illinois	10	–	Gas phase condensation

produced by Nanotek Technology. Powders were uniaxially pressed with 200 MPa to disks of 15 nm diameter and a height of about 2 mm. After pressing, the green density was 48–50% TD. Some powders were dispersed in water and freeze dried before pressing. Hot pressing was performed in graphite dies with polished TZM interlayers under 1100 mbar argon pressure. The samples were then annealed in air for at least 1 h at 1000°C.

Microstructure was analysed by X-ray diffraction. Densities of the samples were measured by Archimedes' method; 5.9 g/ccm was taken as the theoretical density of the cubic zirconia. By X-ray diffraction the constitution of the composites was determined and line broadening was used to calculate the grain size by the Scherrer formula.[8]

Chemical analysis was performed to determine impurity levels of two specimens (ICP–MS, ICP–OES). Microstructure was investigated by scanning electron microscopy (SEM), mainly on fracture surfaces of the composites. The SEM was a JEOL JSM T300 operating at 25 keV. Grain sizes were estimated by Multiplying the grain size from fracture surfaces with a factor of 1.2.[9]

Ionic conductivity for one sample was determined by impedance spectroscopy and for the other samples by four-point electrical resistivity measurements between room temperature and 850°C.

3. RESULTS AND DISCUSSION

In Fig. 1(a) the results of the densification of Tosoh powders by hot pressing are given. Freeze drying of the powders led to lower densities than pressing of the untreated powders. An increase of the pressing pressure from 50 to 110 MPa reduces the necessary pressing temperature by at least 50°C. With this pressure density values close to the theoretical density can be achieved during hot pressing at 1100°C. In Fig. 1(b) the relation between grain size and relative density is given. For the production of fine grained/nanocrystalline samples one has to reach the lower right corner. The grain size values achieved in our samples (<150 nm) are much lower than those achieved by Van de Graaf *et al.*[3] during pressureless sintering. On the other hand the grain sizes of ZrO_2/3 mol% Y_2O_3 prepared by pressureless sintering

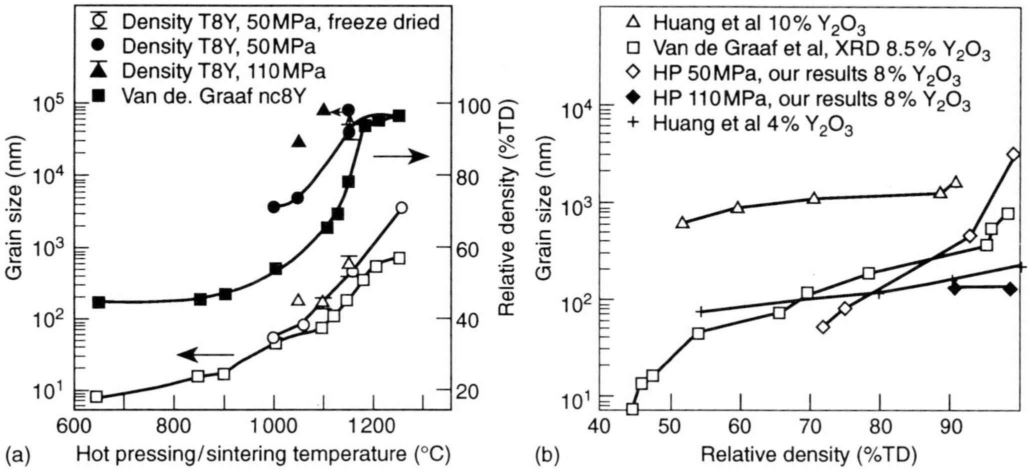

Figure 1(a) Density and grain size of zirconia powders after hot pressing at different temperatures. For comparison the literature results on sintering are also given. (b) Grain size over relative density plot of different zirconia samples.

Table 2
Grain sizes (determined by SEM and by the Scherrer formula (XRD line broadening)) of different hot pressed ZrO_2/8 mol% Y_2O_3 samples

Specimen	Grain size (nm)	
	XRD	SEM
HP 1100°C, annealed 1 h at 1000°C	81 ± 18	110 ± 10
HP 1100°C, annealed 4 h at 1100°C	150 ± 23	160 ± 16
HP 1100°C, annealed 1 h at 1200°C	–	390

are close to our values while the grain sizes of the ZrO_2/8 mol% Y_2O_3 prepared by the same author[4] are much larger. This indicates that fine grained samples with low amounts of Y_2O_3 can be produced much easier than fine grained samples with high amounts of Y_2O_3.

XRD patterns (Fig. 2) reveal only the appearance of the cubic phase. In Table 2 the values of grain size calculated from the Scherrer formula are compared to results from investigations of fracture surfaces. Results from the Scherrer formula are somewhat smaller. This might be due to a difference between grain size and coherent scattering areas, whose size is equal to the influence of internal strains which also contribute to the line broadening.

The results of chemical analysis of hot pressed samples given in Table 3 reveal the

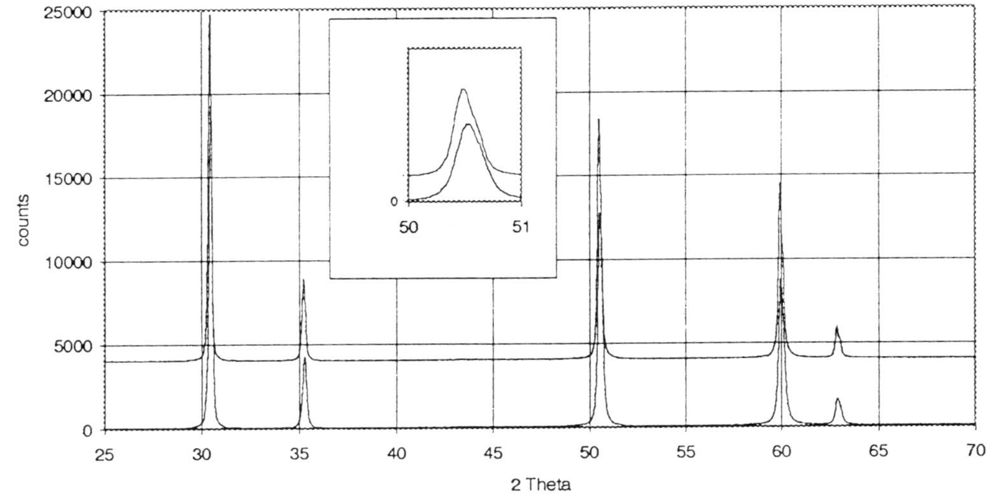

Figure 2 XRD patterns of ZrO_2/8 mol% Y_2O_3 samples. The upper pattern corresponds to a sample annealed for 4 h at 1100°C, the lower to a sample annealed for 1 h at 1000°C.

Table 3
Impurity levels of different ZrO_2/8 mol% Y_2O_3 samples hot pressed at 1100°C

Specimen	Cd	Pb	Al	Be	Zn	Cu	Co	Ni	Ca	Fe	Mg	Na	Mo	Si
4 h at 1100°C in air	<5	<1	60	120	10	<20	<15	40	10	130	<5	190	50	50
1 h at 1000°C in air	<5	<1	60	8	<10	<20	<15	45	20	80	<5	150	70	20

low impurity level of our specimens. The high Be level in one specimen is due to a broken BeO insulation.

In Fig. 3 micrographs of the fracture surface of hot pressed samples are given. The large influence of hot pressing temperature on the final grain size can clearly be seen. Obviously, conventional sintering temperatures for cubic zirconia (>1300°C) lead to grain size in the micrometer range.

Results of impedance spectroscopy measurements performed on the samples with approximately 390 nm grain size between 280 and 600°C are shown in Fig. 4 as Bode-plots. The impedance data could be analysed very well assuming the brick layer model.[10] At low temperatures (200°C) the Bode-plot for the phase shift α shows three well-separated capacitative contributions, which can be assigned to bulk (1 MHz), grain boundary (1 kHz) and electrode (10 mHz). The corresponding Bode-plots for the modulus (z) at 200°C show the two plateaus corresponding to the bulk resistance R_b (around 10 kHz) and the sum of bulk and grain boundary resistances $R_b + R_{gb}$ (around 10–100 Hz). With increasing temperatures the resonance frequencies of all capacitative contributions shift to higher values,

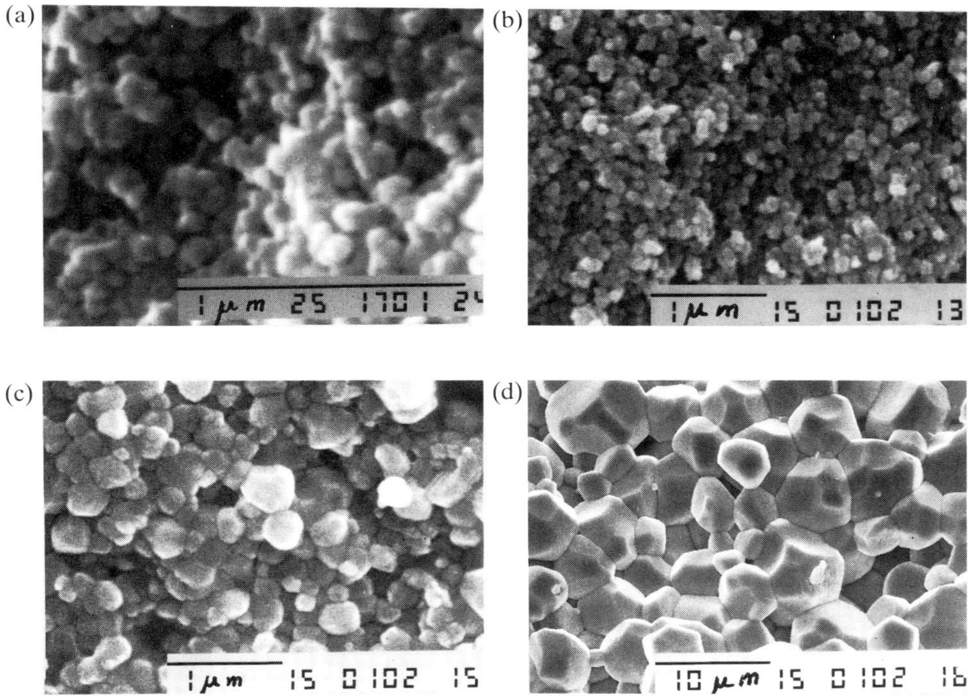

Figure 3 Micrographs of fracture surfaces of ZrO_2/8 mol% Y_2O_3 samples hot pressed at (a) 1000°C, (b) 1050°C, (c) 1150°C and (d) 1250°C.

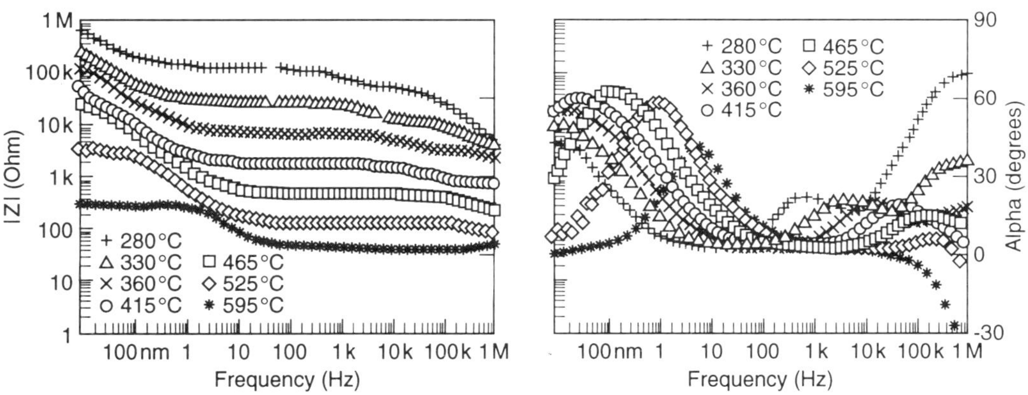

Figure 4 Results of impedance spectroscopy measurement of a ZrO_2/8 mol% Y_2O_3 sample with a grain size of 390 nm.

whereby the contribution of the bulk reaches values outside of the frequency range of the impedance analyser used. Figure 5 shows the Arrhenius plots for the bulk, the grain boundary, and the total conductivity as obtained from the analysis of the impedance spectra using the EQUIVCRT program by B. A. Boukamp. As can be

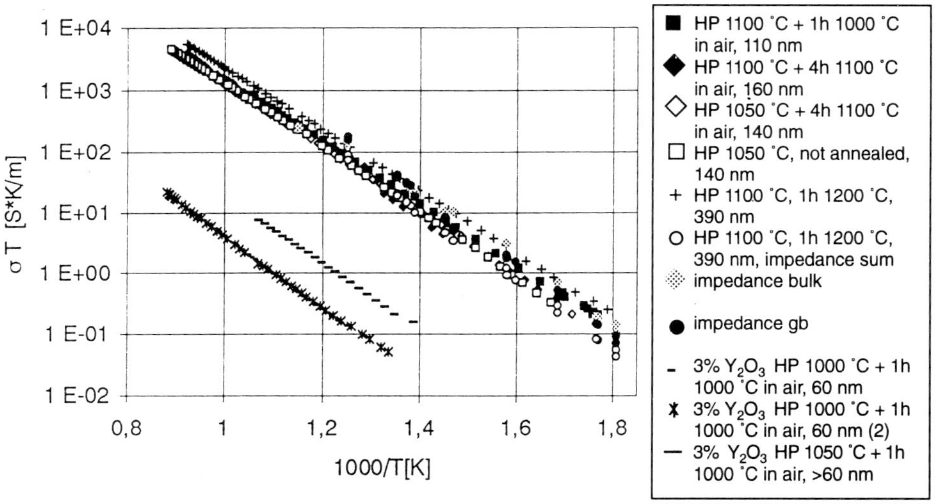

Figure 5 Arrhenius plot of conductivity times temperature for different zirconia samples.

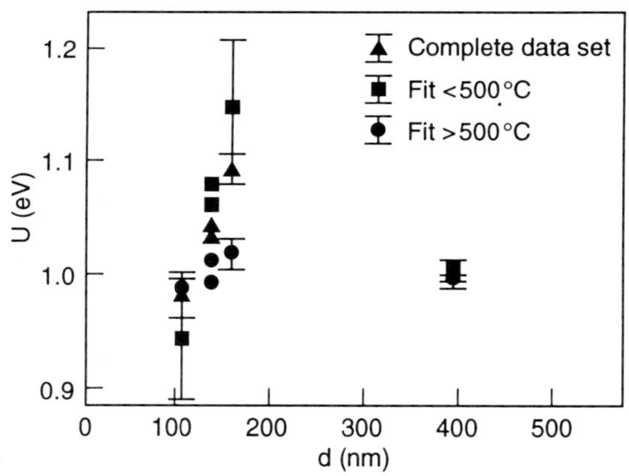

Figure 6 Activation energies of the conductivity times temperature data given in Fig. 5.

seen in Fig. 5, the total conductivity determined from the impedance spectra gives lower values as the results of the four-point conductivity measurements performed on the same sample.

For the fine grained samples only four-point conductivity measurements have been performed; the results are also given in Fig. 5. For comparison, some results from hot pressed ZrO_2/3 mol% Y_2O_3 specimens are also given. Due to the high porosity level of these samples the total conductivity is considerably lower than those of the ZrO_2/8 mol% Y_2O_3 samples.

In order to investigate the effect of grain size on the conductivity least square fits have been performed for the total data sets and for the data points below and above

500°C. The results are plotted in Fig. 6. It was found that in the size range below 200 nm a decrease of activation energy with a decrease of grain size was observed. This is due to a deviation from the Arrhenius behaviour at low temperatures. Especially the fine grain 110 nm sample shows this behaviour.

It was also found for the ZrO_2/3 mol% Y_2O_3 samples with a grain size of about 60 nm which showed an activation energy of about 1.0 eV below and 1.2 eV above 500°C. The decrease in activation energy at low grain size might be a first hint that grain boundary diffusion can play an important role in fine grained zirconia samples to increase conductivity.

The decrease in activation energy at large grain sizes can be explained with results often found in conventional zirconia ceramics.[7,11] It is observed that the grain boundary resistivity, which is aerially connected to the bulk, has a larger activation energy than the bulk resistivity. So with larger grain size the contribution of the grain boundary and also the activation energy decreases.

4. CONCLUSIONS

It is possible to produce dense ZrO_2/8 mol% Y_2O_3 with grain sizes below 150 nm by hot pressing at 1100°C.

Least square fits of the results of four-point conductivity measurements showed that the activation energy of ionic conductivity decreased for grain sizes below 200 nm. The 110 nm (80 nm from XRD) sample showed a deviation from the Arrhenius behaviour at low temperatures. Assuming that this branch is due to grain boundary diffusion, conductivity will exceed the 950°C value of the 400 nm samples if grain size is further reduced at least by a factor of 3 (i.e. 27–50 nm).

REFERENCES

1. C. Suryanarayana: 'Nanocrystalline materials', *Intl Mat. Rev.*, 1995, **40**(2), 41–64.
2. D.-J. Chen and M. J. Mayo: 'Densification and grain growth of ultrafine 3 mol% Y_2O_3-ZrO_2 ceramics', *Nanostruct. Mat.*, 1993, **2**, 469–478.
3. M. A. C. G. Van De Graaf, J. H. H. Ter Maat and A. J. Burggraaf: *Microstructural Development During Pressing and Sintering of Ultrafine Zirconia Powders, Ceramic Powders*, P. Vincenzini (ed.), Elsevier, Amsterdam, 1983, p. 783.
4. K. R. Venkatachari, D. Huang, S. P. Ostranger, W. A. Schulze and G. C. Stangle: 'Preparation of nanocrystalline yttria stabilized zirconia', *J. Mat. Res.*, 1995, **10**(3), 756.
5. C. Chen: 'Fine Grained Zirconia-Metal Dual Phase Composites', Thesis, 1994, University Twente, The Netherlands.
6. M. Mayo, private communications.
7. M. J. Verkerk, B. J. Middelhuis and A. J. Burggraaf: 'Effect of grain boundaries on the conductivity of high-purity ZrO_2-Y_2O_3 ceramics', *Solid State Ionics*, 1982, **6**, 159–170.
8. J. M. Schultz: *Diffraction for Materials Scientists*, Prentice Hall Inc., Englewood Cliffs, New Jersey, vol. 19, p. 226.
9. M. I. Mendelson: 'Average grain size in polycrystalline ceramics', *J. Amer. Ceram. Soc.*, 1969, **52**(8), 443.

10. I. van Dijk and A. J. Burggraaf: *Phys. Status Solidi (A)*, 1981, **63**, 229.
11. H. Kaneko, J. Fuxue and H. Taimatsu: 'Conductivities of yttria-doped zirconia electrolytes', *Proc. of the Third International Symposium on Solid Oxide Fuel Cells*, S. C. Singhal and H. Iwahara (eds), pp. 31–37.

Nature of Electronic Defects in Yttria-stabilised Zirconia and Their Influence on Oxygen Diffusion

R. I. MERINO, N. NICOLOSO and J. MAIER
Max Planck Institut für Festkörerforschung, Heisenbergstrasse 1, D-70569 Stuttgart, Germany

ABSTRACT
In situ optical absorption and diffusion studies have been performed on nominally pure and weakly transition metal-doped yttria-stabilised zirconia single crystals (9.4 mol% Y_2O_3) at temperatures between 950 K and 1100 K and oxygen partial pressures ranging from 10^{-22} to 10^{-16} bar.

Although YSZ is a heavily doped ion conductor, the chemical diffusion coefficient ($\tilde{D}$) does not correspond to the electronic diffusivity but depends on deep level redox-active impurities. Its temperature and oxygen partial pressure dependence can be quantitatively interpreted by a generalised ambipolar diffusion model. Comparison with literature data provides further evidence that the scatter in the values for the chemical diffusion coefficient arises from non-intentional doping.

The unambiguous correlation of $\tilde{D}$ to the Ti impurity content and a single activation energy for $\tilde{D}$ of 1.8 eV in all our samples and also in other samples investigated in the literature point strongly towards Ti^{4+} as the dominant trapping site. The 375 nm absorption band associated with the so-called T-defects observed in EPR is attributed to electrons trapped at Ti sites.

1. INTRODUCTION

Since the discovery of defect-controlled fast oxygen ion conduction in doped zirconia,[1] a great deal of work has been dedicated to the understanding and improvement of its ionic transport properties. Only few contributions have dealt with the identification and characterisation of the minority charge carriers, that is, electrons in the low oxygen partial pressure range, and electron holes at higher oxygen activities.[2–6]

Throughout the main literature it is generally assumed that the chemical diffusion coefficient of an ionic conductor (in which both ionic conductivity and concentration dominates over the respective electronic contributions) is given by the diffusion coefficient of the electronic centres. Therefore, the chemical diffusion coefficient ($\tilde{D}$) of yttria-stabilised zirconia (YSZ) should be independent of additional impurities as long as their concentration is small compared to the yttria doping. This is in sharp contrast to experimental findings. In ref. 7 it has been shown that the influence of redox-active impurities on $\tilde{D}$ can be modelled by the inclusion of source and sink terms due to deep trap levels in the diffusion equation. A special case of these considerations has been anticipated in ref. 8.

Apart from transport studies, a number of spectroscopic investigations of stabilised zirconia have been carried out including optical absorption,[9] photoluminescence[10–12] and electron paramagnetic resonance (EPR).[13–15] In particular, great efforts were made to identify the EPR-active centre in calcia- and yttria-stabilised zirconia and its corresponding optical absorption band at 375 nm which both appear upon reduction of nominally pure materials.[9,13–17] Its origin was attributed successively to F-type centres,[9,16] electrons trapped at a Zr^{4+} site,[13,14,17,18] and a foreign common impurity.[15]

In this work it will be shown that by combining chemical analysis, optical studies and diffusion experiments (i) a clear assignment of the dominant (impurity) trapping site for electrons can be made and (ii) the influence of trapping reactions on the chemical diffusion coefficient can be satisfactorily explained just as in the case of $SrTiO_3$ and oxidised YSZ.[8]

2. EXPERIMENTAL

Polished slices ($A = 10 \times 12\ \mathrm{mm}^2$, $d \leqslant 1$ mm) of transition metal ion doped YSZ single crystals (9.5 mol% Y_2O_3) from ZIRMAT Corporation (MA, USA) have been studied *in situ* by a combined optical absorption and diffusion experiment described in ref. 8. The measurements were carried out at temperatures of $950\ \mathrm{K} < T < 1100\ \mathrm{K}$, and oxygen partial pressures $10^{-22} < p_{O_2} < 10^{-16}$ bar (accessible by mixtures of CO/CO_2 or H_2/CO_2). The actual p_{O_2} is measured down stream with a home-made lambda probe.

Since the impurity content of the samples is assumed to be of crucial importance, we have also performed a precise chemical analysis by proton-induced X-ray emission (PIXE).

3. RESULTS AND DISCUSSION

3.1 Identification of Electron Trapping Sites

The results for the more abundant impurities obtained by PIXE are given in Table 1 and will be presented in more detail elsewhere.[19] Besides the deliberately added dopants (0.1% wt of CoO and Nb_2O_5) and the naturally occurring Hf, the investigated crystals contain appreciable quantities of impurity ions, notably Ti, Ru and Fe, which can also act as trapping centres for electrons.

In accordance with previous optical studies, the oxidised YSZ:Co samples show a single structured absorption band in the visible range due to Co^{2+} in an octahedral environment.[10,20,21] In YSZ:Nb one finds an increased absorption near the fundamental edge. Upon reduction a new band appears in both systems peaking at about 375 nm (3.3 eV and FWHM of 1.35 eV), the same band that is reported in the literature for nominally pure YSZ. Simultaneously, the well-known trigonal paramagnetic resonance signal occurs.[13–15] No additional EPR signal, especially that from Nb^{4+}, can be detected (see Fig. 1). In agreement with this finding, but in contrast to the results of ref. 20, no change in the intensity of the Co^{2+} absorption band is observed. Therefore, we can conclude that the cations Co^{2+} and Nb^{5+} remain in their initial oxidation state upon moderate reduction.

Table 1
Chemical analysis of the YSZ:Co and YSZ:Nb samples (PIXE method, according to ref. 19). Concentration of ions in atppm

	YSZ:Co (380 ppm)	YSZ:Nb (300 ppm)
Hafnium	2650	2250
Ruthenium	345	250
Titanium	140	260
Iron	9	15

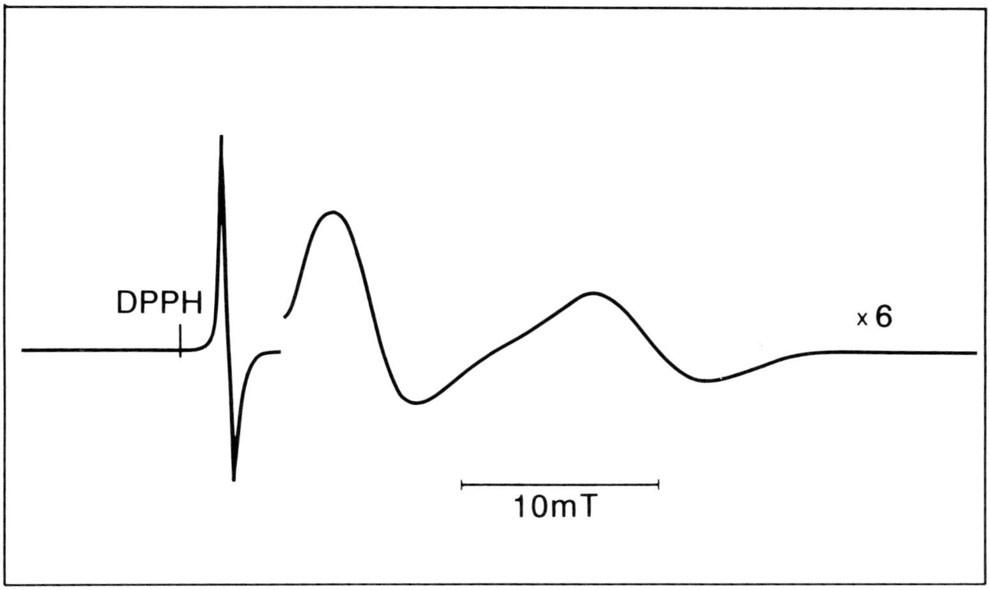

Figure 1 X-band EPR spectrum of the reduced YSZ:Nb sample taken at 77 K with the magnetic field along the ⟨111⟩ direction. The position of the DPPH-marker signal which appears at 331.86 mT is marked with a straight line. The high field signals have been enhanced by a factor of 6. Note that only Ti related signals and no additional signal from Nb^{4+} is observed.

The concentration of trapped electrons ($N_T(T)$) has been calculated from the 375 nm absorption band assuming a value of 0.01 for its oscillator strength.[14,15] For the YSZ:Co sample one obtains

$$N_T(T) = 8.4 \times 10^{19}(\mathrm{cm}^{-3}\,\mathrm{bar}^{1/4})p_{O_2}^{-1/4}\exp\{-1.57(\mathrm{eV})/\mathrm{kT}\}. \tag{1}$$

In Fig. 2 we plot the N_T for four different samples upon a treatment at 900°C in 5%H_2–Ar mixture versus the total Ti content determined by PIXE. The linear dependence and the crossing through the coordinate origin provides clear evidence

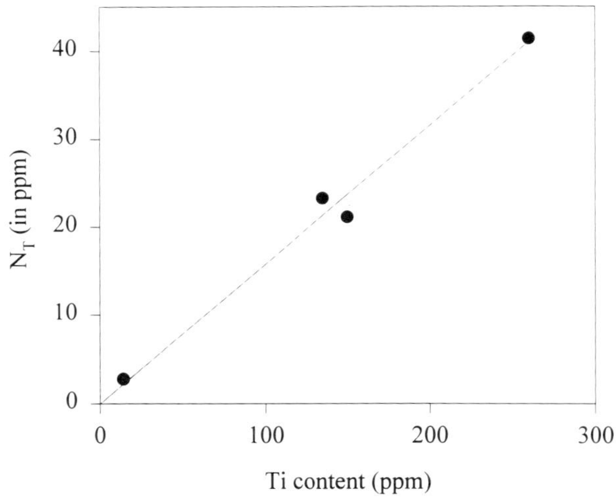

Figure 2 Linear relationship between the overall Ti concentration determined by proton-induced X-ray emission[19] and the concentration of trapped electrons obtained from optical absorption.

for Ti as the dominant trapping site. Recent re-analysis of the paramagnetic resonance has arrived at the same conclusion.[22]

Similarly, a clear correlation exists between the Ti impurity content and the chemical diffusion coefficient ($\tilde{D}$) which has been determined simultaneously by an optical relaxation technique[8] (see Fig. 3). The $\tilde{D}$ values are evaluated from

$$\frac{A(t) - A(\infty)}{A(0) - A(\infty)} = \frac{4}{d\sqrt{\pi}}\sqrt{\tilde{D}t} \tag{2}$$

which is valid for short times t and where d is the thickness of the sample and $A(t)$ stands for the absorption coefficient at time t.[23] The temperature dependence of $\tilde{D}$ is shown in Fig. 4 for two samples having different dopants (YSZ:Co and YSZ:Nb), and, what is more important, different Ti concentrations. Inside the accuracy of our measurements $\tilde{D}$ is oxygen partial pressure independent and for YSZ:Co is given by:

$$\tilde{D} = 160 \exp\{-1.77(\mathrm{eV})/\mathrm{kT}\}\,\mathrm{cm}^2\,\mathrm{s}^{-1}. \tag{3}$$

Furthermore, our calculations show that the difference of $\tilde{D}$ for both samples can be explained by their different Ti concentrations, which will be discussed in more detail below. For both samples we obtained the same activation energy of about 1.8 eV. At this point we must conclude that the same kind of defects are controlling both the diffusion and the concentration of these trapped electrons, namely, electrons trapped at Ti sites.

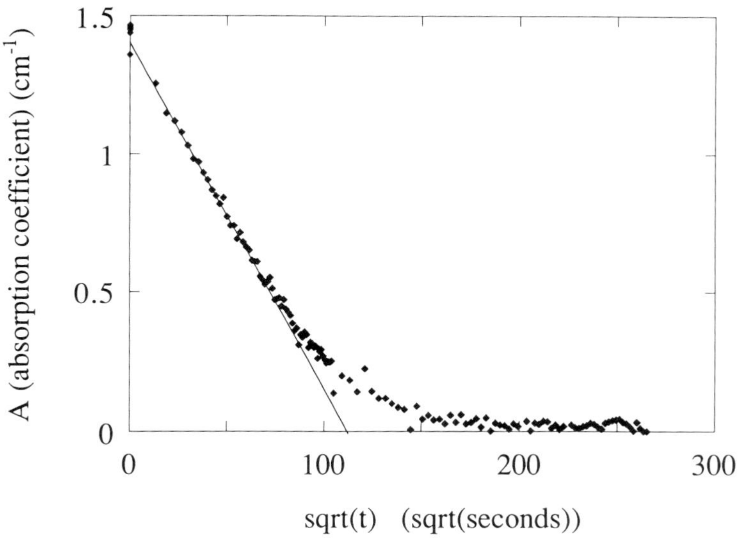

Figure 3 Time variation of the optical absorption coefficient (A) at 405 nm of the YSZ:Co sample upon an oxidation step from 10^{-20} to 10^{-19} bar at 720°C. The straight line shows the $t^{1/2}$ dependence which gives a $\tilde{D} = 1.5 \times 10^{-7}\,\mathrm{cm^2\,s^{-1}}$.

3.2 Influence of Electron Trapping on the Diffusion

In liquid electrolytes and solid electrolytes such as YSZ the diffusion of the ions and electrons is coupled. The chemical diffusion coefficient reduces to

$$\tilde{D} = t_V D_n + t_n D_V \cong t_V D_n \cong D_n \tag{4}$$

since $t_V D_n \gg t_n D_V$ and $t_V \cong 1$ (ie $c_V \gg c_n$ and $\sigma_V \gg \sigma_n$). Here t_j are the transport numbers of conduction electrons (n) and oxygen vacancies (V) and D_j the corresponding diffusion coefficients. In this equation the influence of carrier generation and recombination processes is neglected. It has been shown[7] that for immobile impurities the alteration with respect to eq. (4) consists of an additional factor describing the differential fraction of free (electronic) carriers. Consequently, the result relevant for our situation is

$$\tilde{D} = \chi D_n \tag{5}$$

with the trapping factor (χ) given by

$$\chi = \frac{\delta[\text{free electrons}]}{\delta([\text{free electrons}] + [\text{trapped electrons}])}. \tag{6}$$

This trapping factor, and subsequently $\tilde{D}$, can be calculated from defect chemical relations considering the following local equilibria:

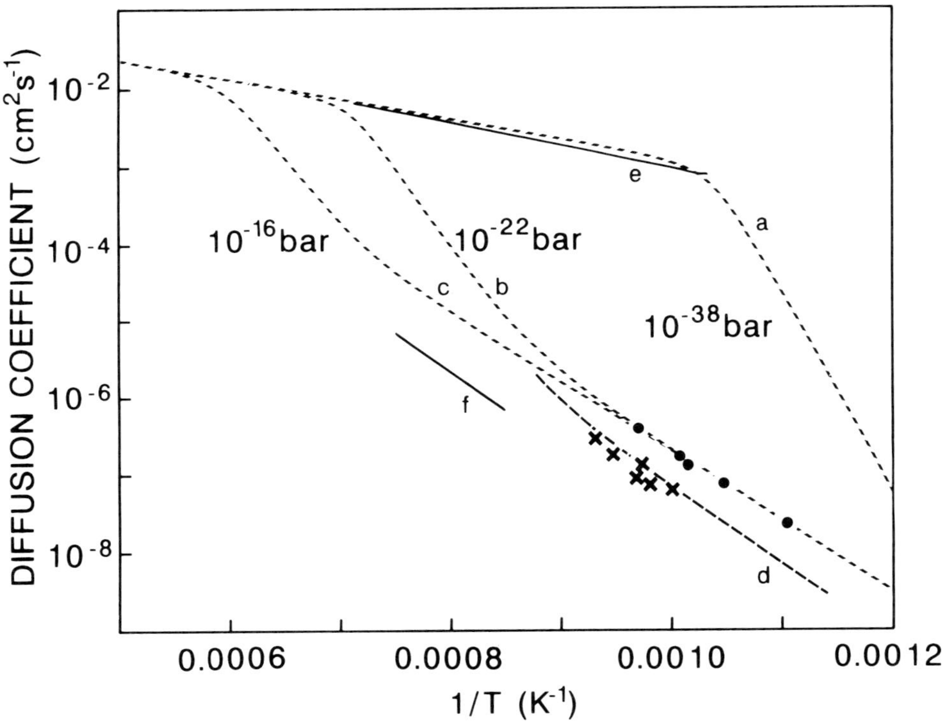

Figure 4 Chemical diffusion coefficient obtained from plots as in Fig. 3 of YSZ:Co (●) and YSZ:Nb (×). It is important to note the different Ti concentrations in both samples (see Table 1). The dashed lines represent calculations using different Ti concentrations (140 ppm for a, b and c and 260 ppm for d) and different oxygen partial pressures (10^{-38} bar (a), 10^{-22} bar (b and d) and 10^{-16} bar (c)). The continuous lines correspond to literature data ((e) polarisation experiments[2,3] and (f) permeation.[6] The electron diffusivity has been taken as $D_n = 0.4\exp(-0.5\,\text{eV/kT})\,\text{cm}^2\,\text{s}^{-1}$, the pre-exponential factors as $K_{20} = 1/N_C = 1.5 \times 10^{-22}\,\text{cm}^3$ and $n_0 = 10^2 N_C = 0.67 \times 10^{24}\,\text{cm}^{-3}\,\text{bar}^{1/4}$.

(i) Oxygen loss, creation of vacancies and free electrons:

$$O_O^X \leftrightarrows \frac{1}{2} O_2(g) + V_o^{\cdot\cdot} + 2e. \tag{7a}$$

As the concentration of oxygen vacancies can be considered constant, one obtains for the concentration of conduction electrons $[e']$

$$[e'] = n_0 p_{O_2}^{-1/4} \exp(-E_{12}/\text{kT}) \tag{7b}$$

where E_{12} is mainly the enthalpy required by the system when one electron is introduced to the edge of the conduction band and 1/4 $O_2(g)$ is extracted to the gas phase. The vacancy concentration has been incorporated in n_0. Note that, since in

YSZ the concentration of oxygen vacancies is fixed by the yttrium content, the electron concentration and thus also the electronic conductivity do not depend on additional impurities.

(ii) Electron trapping:

$$\text{free trap} + 1 \text{ electron} \leftrightarrows \text{trapped electron} \tag{8a}$$

which analogously yields

$$\frac{[\text{trapped electrons}]}{[\text{free traps}]} = [e']K_2 = n_0 K_{20} p_{O_2}^{-1/4} \exp\{-(E_{12} - E_2)/\text{kT}\} \tag{8b}$$

where E_2 is basically the enthalpy required to promote an electron from the trap to the edge of the conduction band.

Combining eqs (6), (7b) and (8b), the trapping factor can be expressed as

$$\chi^{-1}(T, p_{O2}, m) = 1 + \frac{K_2 m}{(1 + K_2[e'])^2} = 1 + \frac{K_{20} m \exp\{E_2/\text{kT}\}}{(1 + n_0 K_{20} p_{O_2}^{-1/4} \exp\{-(E_{12} - E_2)/\text{kT}\})^2} \tag{9}$$

with m being the total amount of traps (free + occupied traps). It is noteworthy that χ and thus $\tilde{D}$ now depend on this quantity.

In the limiting case when the traps are far from being completely occupied, i.e. $K_2[e'] = [\text{trapped electrons}]/[\text{free traps}] \ll 1$, χ and $\tilde{D}$ become oxygen partial pressure independent. As demonstrated in Fig. 4, this condition holds in our experiment. Within this approximation $m \cong [\text{free traps}]$, χ reduces to $(K_2 m)^{-1} = [e']/[\text{trapped electrons}]$, and the electronic conductivity can be correlated with $\tilde{D}$ using the Nernst–Einstein equation

$$\sigma_n = \frac{e^2}{\text{kT}}[e']D_n = \frac{e^2}{\text{kT}}[\text{trapped electrons}]\tilde{D} \tag{10}$$

where D_n is the electron diffusivity.

Table 2 presents the electronic conductivity derived using eqs (1), (3) and (10). For comparison literature values are given. Both our conductivity and activation energy values are in good agreement with the reported data.

In the next step the chemical diffusion coefficient and the concentration of electronic defects were simultaneously fitted in order to demonstrate that the scatter in the literature $\tilde{D}$ values can be quantitatively explained by the trapping effect: under our experimental conditions ($K_2[e'] \ll 1$) we can write

$$[\text{trapped electrons}] = K_2 m[e'] \tag{11a}$$

$$\tilde{D} = D_n/K_2 m = (D_{n0}/K_{20} m)\exp(-(E_D + E_2)/\text{kT}) \tag{11b}$$

Table 2
Comparison of the electronic conductivity of YSZ (from eq. (10), $T = 1000$ K, $p_{O_2} = 10^{-20}$ bar) and its activation energy (E_σ) with literature data

References	$\sigma_n/\Omega^{-1}\,\text{cm}^{-1}$	E_σ/eV
This work	4×10^{-8}	3.4
24	1×10^{-8}	4.1
5	1×10^{-8}	3.7
6	4×10^{-8}	3.9

Equations (11a) and (11b) fix the activation energies $E_{12} - E_2$ and $E_D + E_2$, respectively. In Fig. 5 we have compared the redox constant of the Ti^{3+}/Ti^{4+} equilibrium with the values reported in a previous work where the equilibrium between Fe and Ti dopants in stabilised zirconia has been studied at higher temperatures.[25] The straight line drawn through the data set yields the activation energy $E_{12} - E_2$ (1.8 eV). Using eqs (11b) and (3), we find $E_D + E_2 \cong 1.8$ eV, and thus $E_\sigma = E_{12} + E_D \cong 3.6$ eV, which we use in the following as a more precise value because it takes into account a larger temperature range.

A rough estimate of the pre-exponential factors (K_{20} and n_0) from statistical mechanics yields $n_0 K_{20} \approx 10^{3\pm1}$ bar$^{1/4}$. Taking 10^2 for this product (and using the experimental parameters $E_{12} - E_2 = 1.8$ eV and $E_{12} + E_D = 3.6$ eV) a close fit of the concentration of trapped electrons and the chemical diffusion coefficient is obtained.

In Fig. 4 we present our experimental diffusion data as well as the result of calculations based on eqs (5) and (9), using the parameters given above. We did the calculations for different oxygen partial pressures and the different Ti concentrations which we determined experimentally (Table 1). The agreement between the experimental and calculated curves indicates that the difference in the chemical diffusion data arises from the presence of specific redox active species, here titanium impurities.

In the same figure we also present literature data obtained with different methods (voltage relaxation[2,3] and permeation[6]). Besides experimental difficulties (e.g. affecting the boundary conditions in the polarisation experiment,[26] the large scatter of these $\tilde{D}$ values most likely arises from the chosen experimental conditions (T and p_{O_2}) and non-intentional doping (which makes $\tilde{D}$ oxygen partial pressure dependent). Comparatively small oxygen activities are obtained at the blocking electrode in the polarisation experiments, which in fact are providing the highest experimental values for the chemical diffusion coefficient.[2,3] If we tentatively consider the latter ($\tilde{D} = 0.4\exp\{-0.5\,\text{eV/kT}\}\,\text{cm}^2\,\text{s}^{-1}$) as a lower limit of the diffusivity of the electronic carriers D_n, reliable fits to our experimental values can be obtained provided $N_C \approx 0.5\times10^{22}\,\text{cm}^{-3}$ (N_C being the density of states where the electrons become sufficiently mobile). Such N_C values, together with the non-zero activation energy and the quite low absolute D_n value, may be taken as a first hint for transport in a narrow band created by d-orbitals at the conduction band edge. In view of the

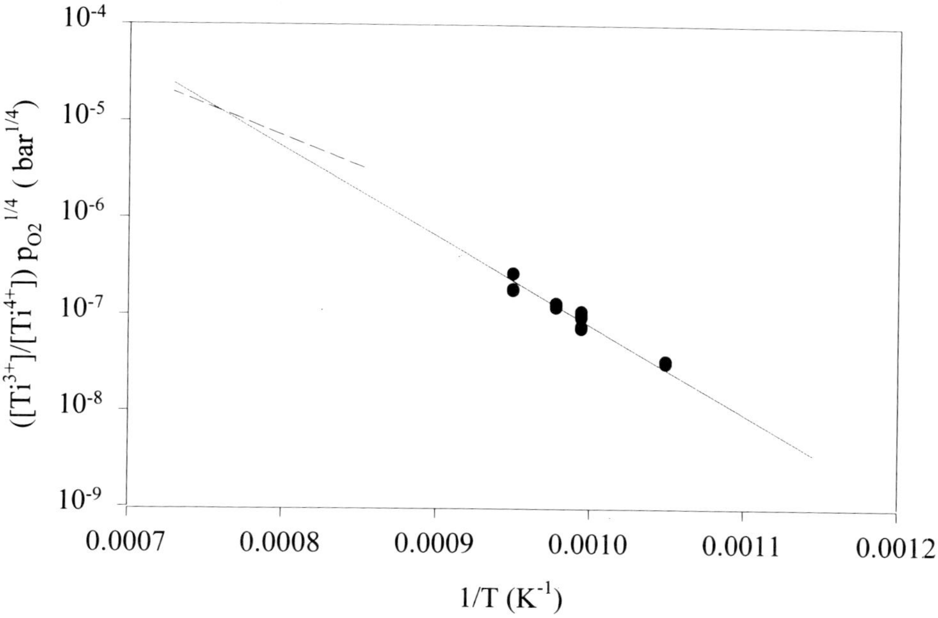

Figure 5. Redox constant $(([Ti^{3+}]/[Ti^{4+}])p_{O_2}^{1/4})$ as a function of 1/T (slope: $E_{12} - E_2 = 1.8$ eV). The high temperature data (dashed line) have been taken from ref. 25.

large uncertainty in the experimental D_n value and the lack of a reliable theoretical estimate, this description has to be looked at with a fair amount of caution. One has to bear in mind that fits of similar quality can be obtained with the same parameter set except $N_C \approx 10^{20}\ cm^{-3}$ and (a non-activated) $D_n \approx 25\ cm^2\ s^{-1}$ which would indicate transport in a wide band. More experimental evidence is required to distinguish between these two kinds of transport mechanisms.

4. SUMMARY AND CONCLUSIONS

In this work we have shown that the optical absorption and the chemical diffusion coefficient of YSZ (n-side) are extrinsically controlled by deep trap levels, most frequently Ti based defects. In our samples, both the intensity of the 3.3 eV optical absorption band and the magnitude of the chemical diffusion coefficient correlate with the Ti content (and only with Ti!) determined by chemical analysis.

In nominally pure YSZ these defects strongly influence the diffusion properties: depending on the impurity concentration (Ti^{3+} or other deep trap levels), the chemical diffusion coefficient may change by orders of magnitude. Its temperature, oxygen partial pressure and impurity concentration dependencies can be explained by a generalised ambipolar transport model and compared with reported data providing clear evidence that the scatter in the literature values arises from non-intentional doping by redox-active centres.

ACKNOWLEDGEMENTS

RIM would like to thank the Dirección General de Investigación Científica y Técnica of the Ministerio de Educación y Ciencia of Spain for financial support. Financial support by the Landesregierung Baden-Württemberg (KKS Program) and technical support by the crystal preparation service group of the Max Planck Institut für Festkörperforschung, Stuttgart, is gratefully acknowledged.

REFERENCES

1. C. Wagner: *Naturwissenschaften*, 1943, **31**, 265.
2. L. D. Burke, J. H. Rickert and R. Steiner: *Z. für Physik. Chem. Neue Folge*, 1971, **74**, 146.
3. W. Weppner: *Z. für Naturforsch.*, 1976, **31a**, 1336.
4. L. Heyne and N. M. Beekmans: *Proc. Brit. Ceram. Soc.*, 1970, **19**, 229.
5. E. Fernandez: thesis, Grenoble, 1980.
6. J. H. Park and R. N. Blumenthal: *J. Electrochem. Soc.*, 1989, **136**, 2867.
7. J. Maier: *J. Am. Ceram. Soc.*, 1993, **76**, 1223.
8. T. Bieger, H. Yugami, N. Nicoloso, J. Maier and R. Waser: *Solid State Ionics*, 1994, **72**, 41. See also T. Bieger, J. Maier and R. Waser: *Sensors and Actuators*, 1992, **B7**, 763.
9. J. S. Thorp, A. Aypar and J. S. Ross: *J. Mater. Sci.*, 1972, **7**, 729.
10. M. Kunz, H. Kretschmann, W. Assmus and C. Klingshirn: *J. Lumin.*, 1987, **37**, 123.
11. E. D. Wachsman, N. Jiang, C. W. Frank, D. M. Mason and D. A. Stevenson: *Appl. Phys.*, 1990, **A50**, 545.
12. S. E. Paje and J. Llopis: *Appl. Phys.*, 1994, **A59**, 569.
13. C. B. Azzoni and A. Paleari: *Phys. Rev.*, 1989, **B40**, 6518 and 9333.
14. V. M. Orera, R. I. Merino, Y. Chen, R. Cases and P. J. Alonso: *Phys. Rev.*, 1990, **B42**, 9782.
15. R. Ben-Michael, D. S. Tannhauser and J. Genossar: *Phys. Rev.*, 1991, **B43**, 7395.
16. J. Genossar and D. S. Tannhauser: *Cryst. Latt. Def. and Amorph. Mat.*, 1987, **16**, 1.
17. C. B. Azzoni and A. Paleari: *Phys. Rev.*, 1991, **B44**, 6858.
18. K. K. Ermakovich, V. N. Lazukin, I. V. Chepeleva and V. I. Aleksandrov: *Sov. Phys. Solid State*, 1976, **18**, 1022.
19. D. W. Plachke, N. Pazarkas, N. Nicoloso and H. D. Carstanjen: To be published.
20. V. I. Aleksandrov, S. Kh. Batygov, M. A. Vishniyakova, Yu. K. Voron'ko, V. F. Kalabuchova, S. V. Lavirschev, E. E. Lomonova, V. A. Myzina and V. V. Osiko: *Sov. Phys. Solid State*, 1984, **26**, 799.
21. V. M. Orera, R. I. Merino, R. Cases and R. Alcalá: *J. Phys. Condens. Matter*, 1993, **5**, 3717.
22. R. I. Merino, V. M. Orera, E. E. Lomonova and S. Kh. Batygov: *Phys. Rev. B*. 1995, **52**, 6150.
23. J. Crank: *The Mathematics of Diffusion*, 2nd edn, Clarendon Press, Oxford, 1975.
24. W. Weppner: *J. Solid State Chem.*, 1977, **20**, 305.
25. T. Reetz, H. Näfe and D. Rettig: 'Advances in ceramics 12' (Ed. Am. Ceram. Soc.), Science and Technology of Zirconia II, 1984, 591.
26. J. Maier: *Z. Phys. Chem. Neue Folge*, 1984, **140**, 191.

Processing and Transport Properties of Double Layer Electrolytes

T. F. PAIS,[1] F. M. B. MARQUES[1] and G. P. WIRTZ[2]
[1] *Ceramics and Glass Engineering Department, University of Aveiro, 3810 Aveiro, Portugal*
[2] *Department of Materials Science and Engineering, University of Illinois at Urbana-Champaign, Urbana, IL 61801, USA*

ABSTRACT
$ZrCl_4$ and YCl_3 were prepared by high temperature reaction of YSZ (yttria-stabilised zirconia) and CCl_4, in the presence of carbon. The metal chlorides produced in this manner were used to grow YSZ films on different dense electrolyte substrates, by EVD (electrochemical vapor deposition), at ambient pressure (1 atm) and 1100°C. The oxygen permeability of two layer electrolyte cells (substrate + film) is related to the film and substrate electrical transport properties and to film thickness, at constant substrate thickness. Estimates for the oxygen permeability of double layer electrolyte cells are compared with effective growth rates of YSZ films deposited on GCO (Gd_2O_3 doped CeO_2), YSZ and TiYSZ (titania doped YSZ) substrates. The relatively small growth rates observed under all these conditions indicate that the film growth rate is not exclusively determined by the substrate electrical properties, although growth rates qualitatively follow the expected dependence on the electrolyte properties.

1. INTRODUCTION

Westinghouse Electric Corp. has developed a combined CVD/EVD technique to obtain thin dense layers of YSZ,[1–3] a well-known high temperature oxygen ion conductor, used in solid oxide fuel cell (SOFC) technology. CVD (chemical vapour deposition) is the initial step, consisting of gas phase reaction of yttrium and zirconium chlorides with an oxidising gas mixture, to form YSZ deposits on a porous substrate. EVD (electrochemical vapour deposition) begins when the pores become blocked by YSZ. Subsequent film growth requires ionic transport of oxygen between the cathode and anode of the local cell. The usual process requires a reactor operating at low pressure (10^{-3}–10^{-2} atm), moderate temperatures (1000–1200°C), and independent $ZrCl_4$ and YCl_3 sublimation beds.

In this work a different solution is suggested for the preparation of the two metal chlorides inside the EVD reactor, namely from the reaction of CCl_4 with YSZ at ambient pressure. It will be shown both theoretically and experimentally that this procedure is feasible. Furthermore, working at ambient pressure requires a simpler experimental setup than in the conventional EVD process. Vacuum systems can be avoided and the procedure is shown to be quite effective for the deposition of a thin layer of YSZ on a dense ceria-based electrolyte. Processing double layer electrolytes is a subject deserving increasing attention when attempting to find electrolytes with ohmic losses compatible with lower SOFC working temperatures.

The deposition of an electrolyte layer on a dense electrolyte substrate can also be used to obtain additional information on the kinetics of film deposition. Thus this work will also focus on the comparison between experimental evidence and theoretical correlation between the transport properties of both electrolytes (substrate and film), cell oxygen electrochemical permeability and film growth rate, aiming at a critical evaluation of the relationship between these processes.

2. THEORY

2.1 *Processing of Metal Chlorides*

According to the theory being presented, CCl_4 can be reacted with zirconium oxide powder at high temperature to produce the corresponding metal chloride. If this reaction occurs in the presence of carbon, the major species present in the gas phase will be $ZrCl_4$ and CO, in addition to the inert carrier gas (assumed to be N_2, to agree with the experimental conditions used in this work). CCl_4, CO_2, Cl_2 and O_2 will be present only in small concentrations, either unreacted (CCl_4) or as reaction products (CO_2, Cl_2, O_2). The overall result from this reaction can be obtained by considering a number of elementary reactions (r_i), their thermodynamic equilibrium constants (K_i), and additional conditions describing the conservation of some species involved in these reactions, as well as the total pressure inside the reactor working chamber. Examples of these elementary reactions are:

$$2Cl_2 + ZrO_2 = ZrCl_4 + O_2 \qquad (r_1)$$

with

$$K_1 = P(O_2) \cdot P(ZrCl_4)/P^2(Cl_2) \qquad (1)$$

and

$$CCl_4 = C + 2Cl_2 \qquad (r_2)$$

with

$$K_2 = P^2(Cl_2)/P(CCl_4). \qquad (2)$$

In these equations $P(i)$ corresponds to the partial pressure of species i. The reaction involving yttrium oxide and chlorine, occurring in parallel with reaction (r_1), will not be considered here in order to minimise the number of relations to be dealt with.

Reaction (r_1) describes the metal chlorination step and reaction (r_2) describes the thermal decomposition of CCl_4. In fact, as the feed gas mixture is heated to about 1000°C before reaching a graphite boat containing the metal oxide, it is believed that most of the CCl_4 is already decomposed before reaching the metal oxide. This means that the sequence of reactions is the opposite of that previously presented.

Finally, the interaction of solid carbon with this system promotes the formation of carbon oxides, because of oxygen released from reaction (r_1):

$$C + O_2 = CO_2 \qquad (r_3)$$

with

$$K_3 = P(CO_2)/P(O_2) \qquad (3)$$

and

$$C + 1/2O_2 = CO \qquad (r_4)$$

with

$$K_4 = P(CO)/P^{1/2}(O_2). \qquad (4)$$

The values for the equilibrium constants for these reactions can be calculated from available thermodynamic data.[4]

The metal chlorination process can also be considered to occur within a closed system with respect to a number of species: all oxygen present in the gas phase (as CO, CO_2 or O_2) must result from the metal oxide converted to chloride (or from the feed gas mixture, if added in a significant amount to the carrier gas); also, all chlorine in the gas phase (as Cl_2, CCl_4 or $ZrCl_4$) should result from CCl_4 in the feed gas mixture. This means that establishing mass balances for these species is straightforward. On combining these mass balances with the equilibrium constants for reactions (r_1) through (r_4) and a total pressure conservation condition, the composition of the gas mixture leaving the metal chlorination boat and reaching the substrate where deposition should occur can be easily estimated. Because of the relevant thermodynamic equilibria involved, this procedure has the added advantage of determining the oxygen partial pressure at the anode side of the cell. This anode working condition will serve as a reference for the present work. The dependence of this anode working condition on other parameters will be further discussed in the next section.

2.2 *Film Deposition*

In the metal oxide deposition process, where the metal chloride reacts with oxygen permeating through the substrate, the thermodynamic conditions differ significantly from the metal chlorination process just described. The deposition process will occur at an oxygen activity higher than the metal chlorination process, without direct contact with solid carbon. This can be understood from the experimental configuration used in our experiments, schematically shown in Fig. 1. In fact, although a graphite ring is used to direct the gas stream over the substrate, the YSZ film deposition occurs at a significant distance from this graphite element, which has no immediate influence in determining the oxygen partial pressure at the

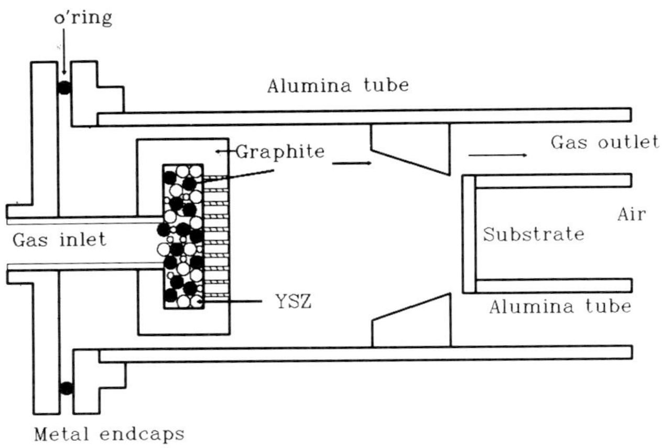

Figure 1 Schematic setup used in the EVD process.

substrate/atmosphere interface. In this case the relevant equilibria involving carbon oxides in the neighbourhood of the substrate is not influenced by the presence of the carbon solid phase, and reduces to:

$$CO + 1/2O_2 = CO_2 \qquad (r_5)$$

with

$$K_5 = P(CO_2)/P^{1/2}(O_2) \cdot P(CO). \qquad (5)$$

Oxygen ions must permeate through the substrate where the film deposition should occur, and this is an essential feature of the EVD process. In the following text the symbol I_o will be used to designate this oxygen ion current flowing through the dense substrate (or film + substrate combination).

Again, to study the operating conditions inside the reactor chamber, a mass balance can be made to estimate the impact of I_o on the partial pressures of the different species in the anode side of the cell. Whether or not such values for I_o can be reached will be determined by the electrochemical permeability of the film and other polarisation losses not considered in this simple type of approach. In any case, this solution can provide a clear view of the interaction between the different parameters involved, namely, the relation between the oxygen partial pressure and the partial pressures of the remaining species. Using the symbols $P^o(i)$ or $P(i)$ to describe the partial pressure of species i in the inlet or outlet molar fluxes, f_i and f_o, respectively, assuming also perfect gas behaviour and that an inert carrier gas (N_2) is used, the mass balance of the relevant species (chlorine, carbon, oxygen and nitrogen) can be described as:

$$[P^o(CCl_4) + P^o(ZrCl_4) + 0.5P^o(Cl_2)] \cdot f_i = [P(ZrCl_4) + P(CCl_4) + 0.5P(Cl_2)] \cdot f_o \qquad (6)$$

$$[P^o(CO) + P^o(CO_2) + P^o(CCl_4)] \cdot f_i = [P(CO) + P(CO_2) + P(CCl_4)] \cdot f_o \qquad (7)$$

$$I_o/4\mathrm{F} = [0.5P(\mathrm{CO}) + P(\mathrm{CO_2}) + P(\mathrm{O_2}) - P(\mathrm{ZrCl_4})]\cdot f_o - [0.5P^o(\mathrm{CO}) + P^o(\mathrm{CO_2}) + P^o(\mathrm{O_2}) - P^o(\mathrm{ZrCl_4})]\cdot f_i \quad (8)$$

$$P^o(\mathrm{N_2})\cdot f_i = P(\mathrm{N_2})\cdot f_o. \quad (9)$$

It can also be assumed that the deposition process occurs at constant pressure. In our experimental setup this is achieved by using a simple gas bubbler at the end of the EVD reactor outlet tube, in direct contact with the atmosphere. This conservation condition can be described as:

$$P^o(\mathrm{N_2}) + P^o(\mathrm{CCl_4}) + P^o(\mathrm{ZrCl_4}) + P^o(\mathrm{Cl_2}) + P^o(\mathrm{CO_2}) + P^o(\mathrm{CO}) + P^o(\mathrm{O_2}) = 1 \quad (10)$$

$$P(\mathrm{N_2}) + P(\mathrm{CCl_4}) + P(\mathrm{ZrCl_4}) + P(\mathrm{Cl_2}) + P(\mathrm{CO_2}) + P(\mathrm{CO}) + P(\mathrm{O_2}) = 1. \quad (11)$$

The analysis of the present situation can be made starting directly from the compositions of the gas mixtures resulting from the metal chlorination process previously described. However, simplified solutions can be assumed to avoid an excessive number of variables to be dealt with. The obvious conditions are:

$$P^o(\mathrm{CCl_4}),\ P^o(\mathrm{Cl_2}),\ P^o(\mathrm{CO_2}),\ P^o(\mathrm{O_2}),\ P(\mathrm{CCl_4}),\ P(\mathrm{O_2}) \approx 0. \quad (c_1)$$

An even simpler solution is to consider the outlet flux identical to the inlet flux, as a result of the carrier gas being the dominant species in the gas phase (typically over 95%):

$$f = f_i = f_o. \quad (12)$$

This procedure reduces the number of variables and correspondingly the number of relations required to solve this problem. Based on this type of approach the following results are obtained:

$$P(\mathrm{ZrCl_4}) = [8K_1P^o(\mathrm{ZrCl_4}) + P(\mathrm{O_2}) - (16K_1P^o(\mathrm{ZrCl_4})P(\mathrm{O_2}) + P(\mathrm{O_2})^2)^{1/2}]/8K_1 \quad (13)$$

$$Io/4Ff = P^o(\mathrm{CO})[1 - 1/2(1 + K_5P^{1/2}(\mathrm{O_2}))] - P(\mathrm{ZrCl_4}) \quad (14)$$

$$P(\mathrm{CO}) = P^o(\mathrm{CO})/2(1 + K_5P^{1/2}(\mathrm{O_2})). \quad (15)$$

For a given value of $P(O_2)$, all the remaining parameters can be obtained by using consecutively eqs (13)–(15), (2) and (5). The results of these types of calculations are shown in Fig. 2 for a working temperature of 1000°C. Full conversion of $ZrCl_4$ to ZrO_2 will only be obtained under oxidising conditions, typically for $P(O_2) > 10^{-8}$ atm. The correlation between I_o and the partial pressures of the different species can be seen in Fig. 3. It is obvious that on adding increasing amounts of oxygen to the system, the first effect is the oxidation of CO to CO_2, and

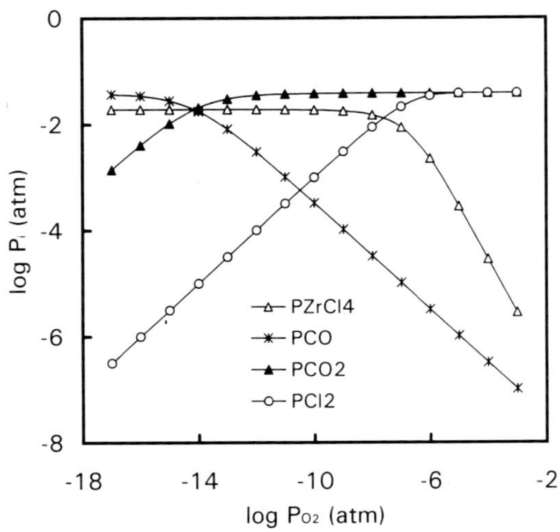

Figure 2 Relation between the partial pressures of dominant species and $P(O_2)$, assuming that the gas mixture fed to the EVD reactor is the result of ZrO_2 conversion to $ZrCl_4$, in the presence of CCl_4 and carbon (N_2 as carrier gas, 2 mol% CCl_4 in the initial gas mixture 1000°C).

only afterwards will the conversion of the metal chloride to oxide effectively begin. The oxygen partial pressure suffers a sharp change near full conversion of CO, which is expected because of the buffering behaviour of reaction (r_5). The same occurs near full conversion of $ZrCl_4$.

2.3 *Oxygen Permeability of Solid Electrolytes*

The permeability of a solid electrolyte like YSZ, expressed as ionic current density (J_o), depends as follows on the working conditions and electrolyte electrical properties:[5]

$$J_o = \sigma_o'(V - V_o)/L \tag{16}$$

where σ_o' is the electrolyte ionic conductivity, L the cell thickness, V_o the ideal nernstian electromotive force of the cell, and V the cell open circuit electromotive force when submitted to a gradient in the oxygen partial pressure ($P(O_2)$). V can be defined as:

$$V = RT/4F \int t_i \, \mathrm{dln}\, P(O_2) \tag{17}$$

with R, T and F having their usual meaning, and t_i being the effective ionic transport number of the electrolyte, corresponding to the ratio between ionic and total conductivity. One useful way to express t_i is:

$$t_i = 1/(1 + K_A \cdot P(O_2)^{-1/4} + K_B \cdot P(O_2)^{1/4}) \tag{18}$$

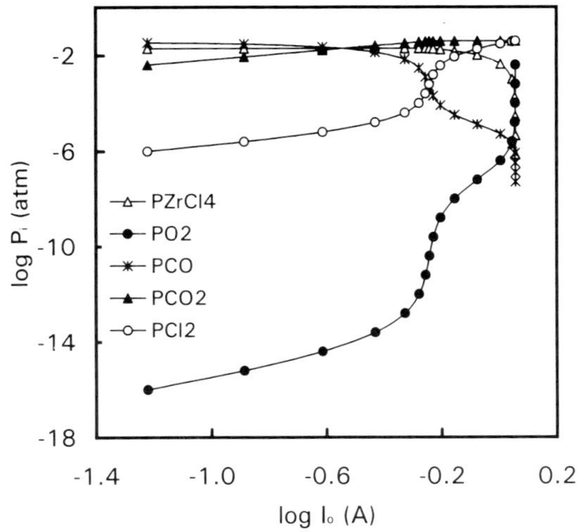

Figure 3 Relation between the partial pressures of different species and I_o, as in Fig. 2.

where K_A is defined as the ratio between the electrolyte *n*-type conductivity and the ionic conductivity at unit oxygen partial pressure. K_B has an equivalent definition in terms of the *p*-type conductivity.

The present work will concentrate on the deposition of YSZ films on dense electrolyte substrates, with various materials being considered as substrates. YSZ will be used as a reference material. Although small differences exist between reported data, there is general agreement on the values reported for ionic and electronic conductivity. Data previously published on the electrical properties of this material will thus be used throughout the present work.[6]

Ceria-based electrolytes have reasonably higher ionic conductivity than YSZ (about double at 1000°C), and a significantly smaller ionic domain due to an increase in the $P(O_2)$ values corresponding to the lower limit of the ionic domain. Exact values for these differences depend on temperature, dopant and dopant concentration. As already mentioned, ceria-based materials are presently the preferred alternative electrolytes for low temperature operation of SOFCs because of their high ionic conductivity. In the present experiments GCO (Gd_2O_3 doped CeO_2) will be used as a substrate. The electrical properties of this material have received special attention in recent years and are also well documented.[7]

Finally, titania-doped YSZ will also be used as a substrate in the present experimental work because of its well-established electrical properties. Namely, marked differences demonstrated with respect to YSZ include a decrease in ionic conductivity (one order of magnitude at 1000°C) and an increase in the lower limit of the ionic domain,[8,9] but not as significant as found for GCO.

Although the selected substrate compositions and properties are reasonably well established, it was decided to use simulated electrical properties derived from YSZ to introduce a more general discussion of the subject of oxygen permeability of

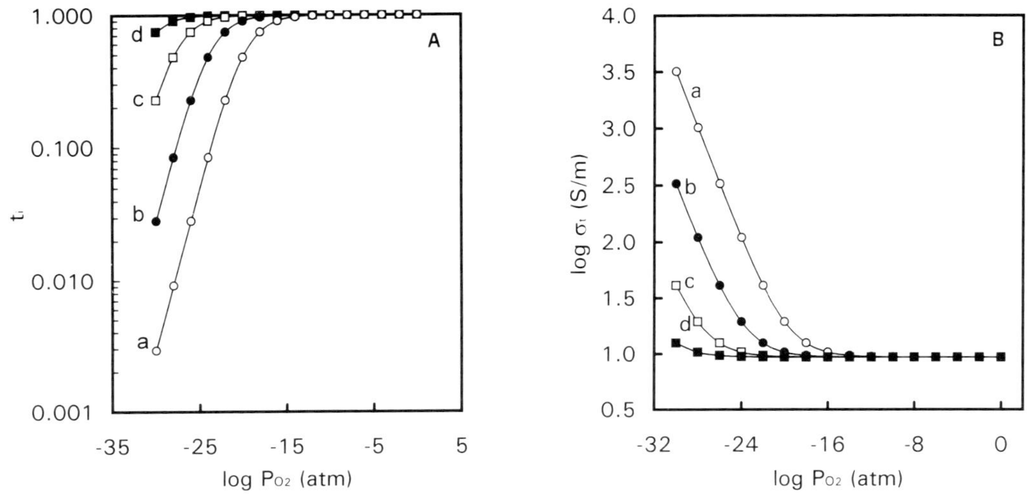

Figure 4 Ionic transport number (A) and conductivity (B) dependence on $P(O_2)$, at 1000°C, for materials considered in the present study (a–1000A, b–100A, c–10A, d–YSZ). For the meaning of symbols, see text.

double layer electrolyte cells. These 'simulated materials', hereby named XA series, are obtained on changing the magnitude of the parameter K_A previously defined in describing the electrolytes' electrical properties (see eq. (18)). In the remaining text an abbreviated notation will be used for this purpose. According to this notation the XA series includes materials where $K_A = 10$ A, 100 A or 1000 A, where A corresponds to the reference value for YSZ. The ionic and *p*-type conductivity of these materials is assumed identical to YSZ. These 'simulated materials' have properties approaching the properties of the substrates being used in these experiments which also makes it possible to comment on the experimental results obtained in this work. Furthermore, in this way it will be possible to show in a systematic manner the role of changing K_A on the oxygen permeability of the double layer electrolyte cells, and on the substrate/film interfacial oxygen pressure under open circuit conditions. This information is quite relevant for the operation of SOFCs, since on going from operating to non-operating conditions the interface between electrolytes will change from high to low oxygen activity.

Figures 4(a), 4(b) and 5 can be used to give a quick view of the characteristics of materials considered for the following discussion. The dependence of the ionic transport number of YSZ and derived materials on oxygen partial pressure is shown in Fig. 4(a). The conductivity diagram for the same materials is shown in Fig. 4(b). It is obvious that such materials differ from YSZ only in the low oxygen pressure behaviour. Finally, Fig. 5 shows the estimated electrolyte electrochemical permeability when one of the cell walls is in contact with air and the other side is brought from oxidising to reducing conditions. Enormous differences can be seen between the permeabilities of such cells. While for the cell anode under moderately oxidising conditions all cell permeabilities tend to be identical, near the plateau

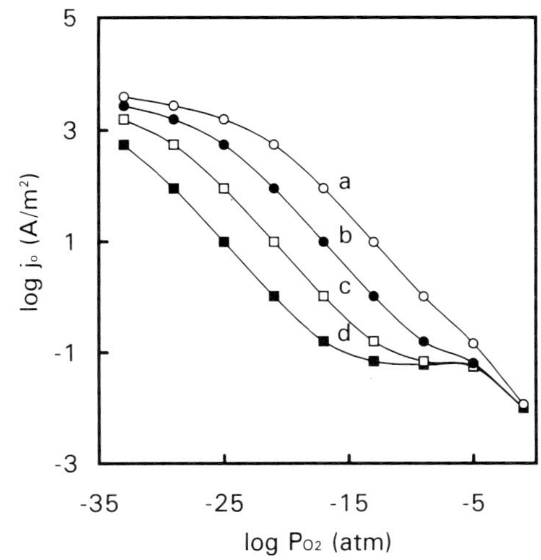

Figure 5 Oxygen permeability of cells 2 mm thick, at 1000°C, with one of the sides in contact with air and the other side in contact with variable $P(O_2)$. Meaning of symbols as in Figure 4.

value observed for YSZ, for the cell anode under typical reducing conditions, the cell permeabilities will differ by orders of magnitude.

2.4 Oxygen Permeability of the Double Layer Cell

A schematic view of the cell and symbols used in the present discussion is shown in Fig. 6. The subsequent examples will always refer to a cell based on a dense electrolyte substrate, with 2 mm thickness (L_1), at 1000°C, with air at the cell cathode side (P_1), a variable oxygen partial pressure at the electrolytes interface (P_2), and an oxygen partial pressure of 10^{-21} atm at the cell anode side (P_3), which corresponds either to the substrate or to the film, if it exists. The film thickness will be represented by L_2. The anode side oxygen pressure boundary condition was selected to correspond closely to the values typical of gas mixtures obtained from reaction between YSZ and CCl_4, at high temperature, in the presence of solid carbon. The substrate electrical properties will correspond to YSZ or one of the materials from the XA series, while the growing film will always be YSZ. All of these conditions are aimed at approaching the effective experimental conditions used in this work.

In the previous section the close relationship between the electrical properties of a given electrolyte and its oxygen permeability has been shown. However, all calculations were made assuming a constant electrolyte thickness. According to our experimental conditions this is true for the underlying substrate but it is not the case for the growing film. Also, the underlying substrate and the growing film usually behave differently, which means that a transition in controlling permeability might occur during film growth, as YSZ will be the material with smaller permeability in

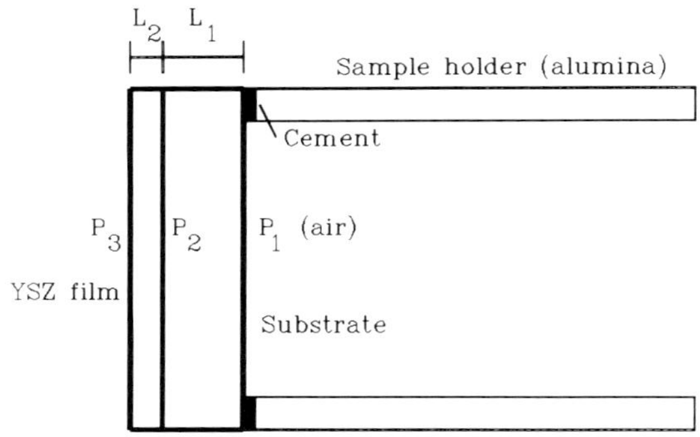

Figure 6 Schematic view of the double layer electrolyte cells considered for the calculations on oxygen permeability and interfacial oxygen partial pressure, shown in Figs 7 and 8. Meaning of symbols as in Figure 4.

all of these examples. As it has not been proven that the film growth rate is determined by the materials' permeability, the discussion will be centred on this aspect. In due time the relationship between oxygen permeability and film growth rate will be analysed in further detail.

To achieve the results shown in Figs 7 and 8 the substrate and film oxygen permeabilities ($J_{o,s}$ and $J_{o,f}$, respectively), were equated:

$$J_{o,s} = J_{o,f}. \tag{19}$$

This corresponds to the steady state ionic current conservation condition in the substrate + film combination. After incorporating eqs (17) and (18) in eq. (19), with T, P_1, P_3, L_1 and L_2 fixed, P_2 was found by iteration for each double layer electrolyte combination.

Figure 7 shows the role of the deposited YSZ film thickness on the cell permeability, for YSZ and XA electrolytes as dense substrates. As would be expected, for negligible film thickness, the substrate controls the oxygen permeation rate, which increases with increasing electronic conductivity of the substrate. For a given film thickness it can be noticed the increasing influence of the deposited film in determining the overall cell permeability. This is indicated by a significant change in slope of the curve representing the oxygen permeability. Furthermore, all curves approach each other for large film thicknesses. This is understandable, keeping in mind that YSZ has the lowest permeability and that this analysis is carried out to a deposited film thickness of 2 mm (equal to the thickness of the substrate). One important observation from this figure is that the oxygen permeability of the different substrates increases by almost three orders of magnitude with K_A ranging from A to 1000 A. It can be also concluded that the film will only have any relevance in determining the permeability of the substrate + film combination when the film thickness exceeds roughly 1 μm (substrate 1000 A), 10 μm (substrate 100 A) or 100 μm (substrate 10 A). However, taking the first case as an example approaching ceria-based electrolyte characteristics, a film 100 μm thick would be required to

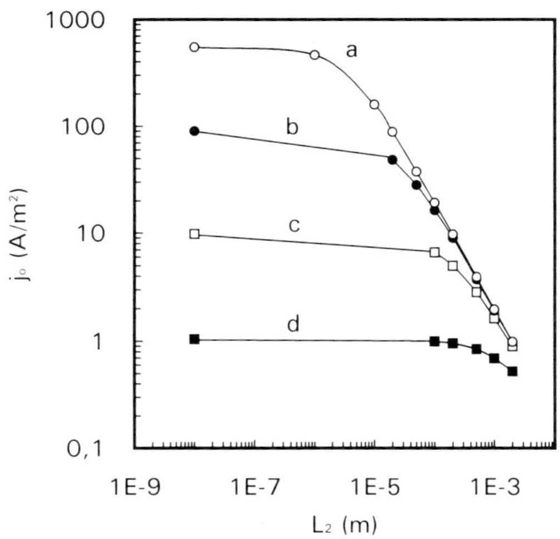

Figure 7 Oxygen permeability of double layer electrolytes (expressed as current density J_o), as a function of the thickness of the YSZ deposited film. Meaning of symbols as in Figure 4.

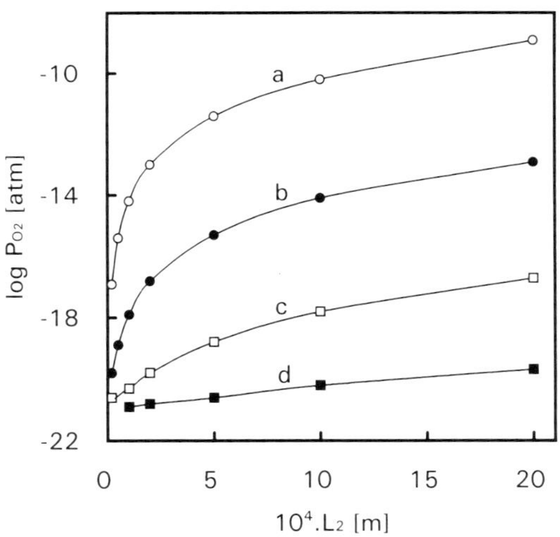

Figure 8 Dependence of the oxygen partial pressure (P_2) at the double layer electrolyte interface, on the YSZ deposited film thickness (L_2). Meaning of symbols as in Figure 4.

lower the cell permeability to about 10 times the permeability of the YSZ substrate.

2.5 Interfacial Oxygen Partial Pressure

The role of the film thickness in changing the oxygen partial pressure at the substrate/film interface (P_2) is shown in Fig. 8 for the YSZ and XA based double layer cells. The higher the substrate permeability the lower the film thickness required to drive the interface to more oxidising conditions. As expected, for low permeability substrates the deposited film is very ineffective. However, Fig. 8 also shows that for a high permeability substrate (like ceria-based electrolytes) a film thickness of almost 100 μm might be required to change P_2 by six orders of magnitude at this temperature. This is especially relevant for SOFCs as the cell voltage drop will increase with decreasing P_2 when ceria-based electrolytes are used. At lower temperatures or under operating conditions, the requirements can be expected to be less severe.

2.6 Oxygen Permeability and Film Growth Rate

Until now only the relationships between the electrolyte properties, working conditions and cell permeability have been considered. The next question to be addressed is the impact of the ionic current flowing through the cell in terms of film growth rate. By assuming that all oxygen permeating through the cell is combined with the metal chlorides to produce the YSZ film, the film growth rate can be easily related to the cell oxygen permeability. Every two moles of oxygen ions permeating through the cell can be converted into one mole of ZrO_2 (i.e. 1 mole of O_2 corresponds to 4.F Coulombs transported as ionic current). Depending on the yttrium content in YSZ, identical estimates can be made for the conversion of oxygen permeability in the oxide solid solution formation. The first conclusion from such estimates is that even for a rather thick substrate, at low temperatures, the film growth rate should be considerably higher than that reported for combined CVD/EVD experiments. The second conclusion is that for substrates with significantly different electrical properties, as those being studied, the film growth rates should differ by orders of magnitude. This is a direct consequence of results shown in Fig. 7, assuming cells under identical cathode and anode working conditions. Considering the existing experimental evidence as well as data being reported in this work, there is good reason to question the efficiency of the conversion of all oxygen permeating through the cell in forming the growing film, or on the usually assumed film growth kinetics controlled by the electrolyte *n*-type conductivity. Additional experimental evidence will be provided in the next sections.

3. EXPERIMENTAL SETUP

An EVD reactor was built based on a horizontal impervious alumina tube connected to a gas feed system capable of providing different dosages of N_2 and CCl_4. The CCl_4 content in the carrier gas (N_2) is fixed by controlling the temperature of a CCl_4 bath through which the carrier gas is forced. The tube reactor includes a graphite boat filled with YSZ powder and graphite grains. The feed gas mixture is brought into contact with this powder in order to promote the reaction between CCl_4 and the oxide solid solution. The resulting streaming gas mixture is then brought into contact with the dense substrate (supported on an inner alumina tube) upon which the EVD deposition is expected to occur. The substrate is sealed against the alumina tube with a gas-tight refractory cement (Fig. 1).

A series of depositions was performed at 1100°C on YSZ, TiYSZ and GCO substrates, for periods of 6 h. The reactor chamber was always kept at a working pressure of 1 atm, with air in direct contact with the cell cathode. The dense substrates used in this work were prepared directly from Tosoh YSZ powder (8 mol% Y_2O_3), from a derived composition with an addition of 10 mol% TiO_2 (TiYSZ), and from GCO powder ($Ce_{0.9}Gd_{0.1}O_{4-\delta}$ from SSC). The powders were uniaxially pressed and sintered at 1600°C for 4 h, to achieve high densification (>95% of theoretical density). This was confirmed from SEM microstructural observations. The substrates were polished with diamond paste to obtain a better definition of the interface between substrate and film. All YSZ films were deposited on dense electrolyte substrates of about 2 mm thickness and 20 mm diameter.

4. RESULTS AND DISCUSSION

YSZ films deposited by EVD were characterised by SEM/EDS and XRD in order to measure the film thickness and also to study the film morphology, composition and structure. For all films it was concluded that they consisted of cubic zirconia, either with a slight excess or deficit in Y when compared to the composition of the starting powder.

The interfaces between film and electrolyte were carefully analysed in all cases. While for films deposited on YSZ and TiYSZ substrates all interfaces seemed perfect with full contact between substrate and film; in the case of ceria-based electrolytes, although in general the adherence between film and substrate was excellent, at some points the film showed peeling, suggesting that a crack had developed at this interface. Typical microstructures observed in all three cases are shown in Fig. 9. A possible explanation for the peeling observed with ceria-based electrolytes is that the deposition process occurred while part of the cerium ions was reduced at the anode side. During subsequent reoxidation, the cerium ions would reoxidise and the substrate would shrink. Since the cations in YSZ are more stable to reduction than cerium ions, a redox cycle would induce an expansion mismatch between the two materials. If this mismatch exceeds some critical value, cracks would develop in the interface between electrolytes. As the operating conditions during EVD are more severe than expected for SOFCs operating at low temperature, it is believed that this effect will not be observed under normal operating conditions. Film deposition conditions can also be optimised to avoid this effect, for example lowering the deposition temperature.

The experimentally observed growth rates were higher for GCO based cells and decreased for TiYSZ and YSZ. In fact the maximum YSZ film thicknesses observed after 6 h of deposition at 1100°C were about 40, 19 and 4 μm, for GCO, TiYSZ and YSZ substrates, respectively. As the three substrate materials being used have electrical properties ranging from pure electrolytic behaviour under reducing

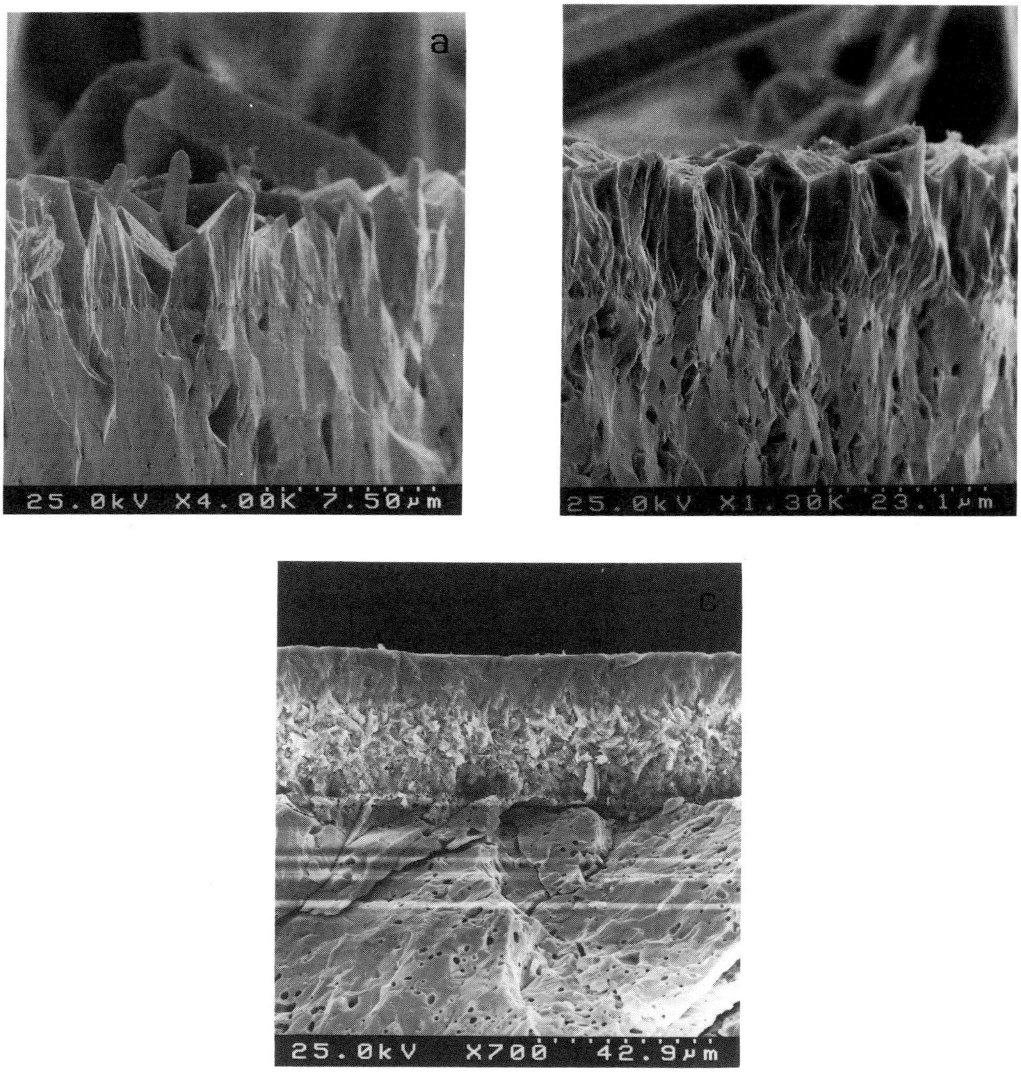

Figure 9 Cross-section views of YSZ films deposited at 1100°C, for 6 h, on (a) YSZ, (b) TiYSZ and (c) GCO dense substrates, 2 mm thick.

conditions (YSZ, with $t_i > 0.99$) to clear mixed conductivity (TiYSZ and GCO, with $t_i \ll 0.99$), in the latter cases the film growth rate should be enhanced by orders of magnitude compared to depositions on YSZ substrates. In fact, unless the film achieves a considerable thickness, the substrate is believed to control the cell permeability. Even assuming some uncertainty for the *p*-type conductivity of ceria-based materials, growth rates differing by about three orders of magnitude should be observed at 1100°C for YSZ films deposited on YSZ and GCO substrates. This suggests that the YSZ film growth rate is not only determined by the combined electrochemical permeability of the film + substrate, but that other phenomena involved in EVD are also likely to play an important role on film growth. The deviation of the present experimental results from predictions based on previously assumed growth rates and mechanisms might be explained by possible side (competitive) reactions involving oxygen consumption, or elementary steps in the process which cause polarisation losses and consequently a voltage drop driving the electronic current through the cell. It seems thus advisable to adopt a rather flexible approach to discuss the role of the different elementary steps involved in the overall reaction, instead of searching for a rate determining step. This can be done by discussing the relation between the current density and film growth rate in terms of polarisation components. The ionic current crossing the cell has already been expressed as a function of the cell voltage, and the same can be done for the remaining elementary steps, at least in a qualitative manner.

A possible competitive reaction for oxygen consumption inside the EVD reactor chamber can be readily identified based on the composition of the reactor feed gas mixture. It has been shown that the composition of the gas mixture entering the reactor deposition zone should include an important concentration of CO because of the processing route used in the formation of the metal chlorides. CO is expected to become almost fully oxidised before any significant conversion of metal chlorides to oxides occurs. This means that, in practice, the oxygen permeating through the cell is being consumed principally in oxidising CO to CO_2 rather than in converting metal chlorides to oxides (see Fig. 3). The effective ratio between oxygen being consumed in CO oxidation and metal chloride oxidation (in the absence of CO) will be determined by the working conditions inside the reactor chamber. This ratio can be estimated from the previously presented mass balance approach. The result is that the ratio between 'inefficient' and 'efficient' utilisation of oxygen can reach more than three orders of magnitude if the reactor operates under reducing conditions, which corresponds to a small conversion of metal chlorides. In these experiments a large excess of metal chlorides was always present. The excess chloride could be collected at the cold end of the furnace tube and identified as a water-soluble chloride, using the traditional $AgNO_3$ analytical test. This suggests that for the present working conditions the level of metal chloride conversion was always quite small and most of the oxygen permeating through the cell was indeed used in CO oxidation rather than in YSZ film formation. This can explain to a large extent the difference between the growth rates experimentally observed and those which might be expected from the electrical transport properties of the film and substrate.

While the previously presented argument can be used in favour of the thesis that film growth kinetics is governed by electron transport, it becomes rather more difficult to accept this argument when comparing results mentioned in the literature on growth rates observed for experimental conditions where only metal chloride sublimation beds and inert carrier gases were used.[10–14] Under such conditions the oxygen transported through the cell would be expected to combine with the metal chlorides to form the YSZ film, and no side reactions can be assumed because of the composition of the feeding gas mixture. In such conditions, for films of relatively small thicknesses without dense underlying substrates, the reported growth rates were only slightly higher than those herein reported. This requires a critical evaluation of other steps involved in the EVD process.

Considering the role of anode polarisation in cell behaviour, it can be readily shown that even small voltage drops will always originate a sharp decrease in the magnitude of the permeating current as seen in Fig. 5. If the cell ever becomes polarised on the anode side, the most likely result would be a drop in oxygen permeability to a value near the plateau shown in Fig. 5 for YSZ. In fact, the best anode working conditions are not far from this situation. In such cases, materials which differ mostly in the magnitude of their *n*-type conductivity should tend to exhibit behaviour similar to YSZ, as all their permeabilities tend to this plateau value at moderately reducing conditions. This subject has hardly received any attention in previous work because most previous work has concentrated on the combined CVD/EVD of YSZ, and differences arising from the use of dense substrates would not be observed. The relatively small range of growth rates observed in this case clearly suggests that this is a likely hypothesis, although the origin of such polarisation phenomena still remains to be identified.

In most experimental conditions the total concentration of metal chlorides in the gas phase is quite small as a result of the low operating pressure, but also because the metal chlorides are so diluted by the carrier gas. However, the EVD reaction requires both gas phase transport of the metal chlorides to the film/gas interface and transport of chlorine away from this interface. In both cases a concentration gradient is necessary to ensure this transport. Significant gas phase concentration gradients of both species are impossible unless moderately oxidising conditions develop at the film/gas interface while the furnace bulk atmosphere composition is kept close to the values resulting from the chlorination process. In fact, it can be seen in Fig. 2 that the concentration of Cl_2 is negligible under reducing conditions, while the concentration of $ZrCl_4$ is almost constant for the same range of conditions. This means that significant gas phase transport rates for reactants and reaction products can only be achieved at the expense of an anodic concentration polarisation.

Little attention has been given to the cathodic process. In this case the reactant is present in large excess in the gas phase (oxygen in air) and concentration polarisation is unlikely to be relevant. However, the required adsorption of oxygen at the electrolyte surface might be important in terms of kinetics. The catalytic activity of the electrolyte surfaces is expected to be low. Significant concentrations of adsorbed oxygen (in equilibrium with the gas phase) can only be assumed in the presence of a catalyst such as one of the commonly used electrode materials. This means that the deposition of a catalytically active material (Pt or a perovskite such as $(La,Sr)MnO_3$) on the cell cathode side should provide some evidence for the role of this surface process in the overall EVD kinetics. One order of magnitude deviation between the oxygen activity in air and at the cell surface (cathode side) would produce a significant abatement in the oxygen permeability of the cell.

The only elementary steps not taken into consideration up to now were the charge transfer processes. These processes are also likely to introduce some polarisation losses in the overall EVD process, but because of the relatively small current densities involved it seems unlikely that they would correspond to the dominant contribution. Finally, it should be remembered that the dependence of film growth rate on substrate composition is, in fact, in qualitative agreement with the cell oxygen permeability dependence on this variable. This means that oxygen transport through the cell unquestionably plays an important role in the overall EVD mechanism.

5. CONCLUSIONS

It has been shown that layers of YSZ can be deposited on different dense substrates by EVD, and that such layers might become quite effective in blocking electron transport through double layer electrolytes, including ceria-based materials. However, the permeability of such materials combinations is far higher than the YSZ film growth rate. This fact can be partly due to the existence of competitive reactions for oxygen consumption inside the EVD reactor. A qualitative analysis of potential contributions to cell polarisation suggests that adsorption reactions might be relevant in the cathode side while concentration polarisation might dominate in the cell anode side. Finally, while the quantitative correlation between the electrolyte properties and film growth rate was not confirmed, qualitative agreement exists between experimental evidence and the underlying substrate electrical properties, which indicates that this factor is nevertheless significant in film growth kinetics.

ACKNOWLEDGEMENT

Work performed with financial support from JNICT (Portugal), contract PBIC/C/CTM/1384/92. F. Marques also wishes to express his gratitude to Dr Michel Kleitz for making possible the research visit to LIESG, France, during which this manuscript was partly prepared.

REFERENCES

1. A. O. Isenberg: 'Growth of refractory oxide layers by electrochemical vapor deposition (EVD) at elevated temperatures', in *Electrode Materials and Processes for Energy Conversion and Storage*, PV 77-6, 572, J. D. E. McIntyre, S. Srinivasan and F. G. Will (eds), The Electrochemical Society Softbound Proc. Series, 1977, Princeton, NJ.
2. A. O. Isenberg: 'Energy conversion via solid oxide electrochemical cells at high temperature', *Solid State Ionics*, 1981, **3/4**, 431.
3. U. B. Pal and S. C. Singhal: 'Electrochemical vapor deposition of yttria-stabilized zirconia films', *J. Electrochem. Soc.*, 1990, **137**(9), 2937–2941.
4. G. P. Wirtz: 'Electrochemical vapor deposition of oxide films', *Morris E. Fine Symposium*, P. K. Liaw, J. R. Weertman, H. L. Marcus and J. S. Santer (eds), The Mineral Metals and Materials Society, Warrendale, PA, 1991, pp. 221–230.
5. F. M. B. Marques and G. P. Wirtz: 'Oxygen fugacity control in non-flowing gas phase systems. I – Experimental results', *J. Am. Ceram. Soc.*, 1992, **75**(2), 369–374.

6. M. Kleitz, E. Fernandez, J. Fouletier and P. Fabry: 'Determination of electronic conductivities and ionic domains of ZrO_2-Y_2O_3 by semipermeability measurements', in *Advances in Ceramics*, vol. 3, 349, A. Heuer and L. W. Hobbs (eds), The American Ceramics Society, 1981, Columbus, OH.
7. M. Mogensen, T. Lindegaard, U. Hansen and G. Mogensen: 'Physical properties of mixed conductor solid oxide fuel cell anodes of doped CeO_2', *J. Electrochem. Soc.*, 1994, **141**(8), 2122–2128.
8. R. M. C. Marques, F. M. B. Marques and J. R. Frade: 'Characterization of mixed conductors by DC techniques. Part II: Experimental results', *Solid State Ionics*, 1994, **73**, 27–34.
9. M. T. Colomer, L. S. M. Traqueia, J. R. Jurado and F. M. B. Marques: 'Role of grain boundaries on the electrical properties of titania doped yttria stabilized zirconia', *Mat. Res. Bull.*, 1995, **30**(4), 515–522.
10. M. F. Carolan and J. N. Michaels: 'Chemical vapor deposition of stabilized zirconia on porous supports', *Solid State Ionics*, 1987, **25**, 207–216.
11. Y. S. Lin, L. G. J. de Haart, K. J. de Vries and A. J. Burggraaf: 'A kinetic study of the electrochemical vapor deposition of solid oxide electrolyte films on porous substrates', *J. Electrochem. Soc.*, 1990, **137**(12), 3960–3966.
12. M. F. Carolan and J. N. Michaels: 'Growth rates and mechanism of electrochemical vapor deposited YSZ films', *Solid State Ionics*, 1990, **37**, 189–195.
13. J. P. Dekker, N. J. Kiwiet and J. Schoonman, 'Electrochemical vapor deposition of SOFC components', in *Proc. of the First International Symp. on Solid Oxide Fuel Cells*, PV 89-11, S. C. Singhal (ed.), The Electrochemical Society Inc., Pennington, NJ, 1989, pp. 57–66.
14. L. G. J. de Haart, Y. S. Lin, K. J. de Vries and A. J. Burggraaf: 'On the kinetic study of electrochemical vapor deposition', *Solid State Ionics*, 1991, **47**, 331.

Single and Multiple Grain Boundary Diffusion Through Membranes

W. PREIS and W. SITTE
Institut für Physikalische und Theoretische Chemie, Technische Universität Graz, A - 8010 Graz, Austria

ABSTRACT
A detailed mathematical analysis of grain boundary diffusion through thin membranes of finite thickness has been performed. The membrane is bounded by two permeable and parallel surface planes. The diffusing particles flow into the solid through one surface and at the opposite surface the mobile particles diffuse out of the solid. A constant flux through these surfaces may be maintained during the entire diffusion process. Fick's second law of diffusion has been solved for an isolated grain boundary as well as multiple grain boundaries, i.e. an array of equidistant parallel grain boundaries. The former case has been solved by applying Fourier–Laplace transforms whereas the analytical solution of the latter case has been obtained by Fourier analysis techniques. Several concentration profile plots are given to illustrate the solutions for both models. Average concentration depth profiles for multiple grain boundaries are compared with those obtained from the isolated grain boundary model. A satisfactory coincidence between the concentration profiles of these two models is achieved when the grain boundary distance is sufficiently high or the diffusion time fairly small.

1. INTRODUCTION

Diffusion along grain boundaries ('short circuit diffusion') in polycrystalline materials proceeds by several orders of magnitude more rapidly than bulk lattice diffusion. Grain boundary diffusion plays an important role in all diffusion-controlled microstructural changes of polycrystalline materials such as sintering, grain growth, recrystallisation, phase transformations and solid state reactions (like e.g. corrosion).[1] Furthermore, polycrystalline thin films have gained great technological importance in large-scale integrated microelectronic devices in recent years. Low temperature diffusion phenomena strongly influence the reliability and lifetime of these devices. In addition, electrochemical applications of thin films can be found in sensors or electrochromic displays or as fast electrodes in miniaturised solid-state batteries. Ionic (oxygen) and mixed ionic–electronic conducting ceramic membranes are part of solid oxide fuel cells and ceramic oxygen generators/separators.[2]

The mathematical treatment of grain boundary diffusion is based on contributions of Fisher,[3] Whipple,[4] Suzuoka[5] and Le Claire.[6] In all these early works the polycrystalline medium has been treated as a semi-infinite solid where the isolated grain boundary is a highly diffusive slab imbedded in the solid. Gilmer and Farrell[7,8] have investigated grain boundary diffusion in thin films of finite thickness. They have considered both isolated and multiple grain boundaries. In all these investigations one surface of the polycrystalline solid is in contact with a constant or instantaneous source. The opposite surface has been treated as a reflective one in the case of finite thickness geometry. Hwang and Balluffi[9] have performed an analysis for the surface accumulation method applied to determine grain boundary diffusion coefficients of thin films by Auger spectroscopy. Among others Gupta *et al.*[10] and Peterson[11] have reviewed grain boundary diffusion. Recently, Gui *et al.*[12] have introduced a transmission line matrix method in order to simulate grain boundary diffusion processes. In addition, Yao and Cahoon[13] have developed an analytical solution for diffusion along cylindrical grain boundaries imbedded in a thin film.

In the past the lack of precise grain boundary diffusion data posed a problem. Much progress has been made by the introduction of the $^{18}O/^{16}O$-gas/solid exchange technique in combination with accurate ^{18}O-depth profile analysis applying secondary ion mass spectrometry (SIMS).[14,15] The parameters available from this technique are the surface exchange coefficient and the tracer-(self)-diffusion coefficient of oxygen.

Our interest in grain-boundary diffusion originates from polarisation experiments on mixed conductors in symmetric solid-state galvanic cells in order to study the kinetic properties of these materials (chemical and component diffusion coefficient, ionic and electronic conductivity).[16,17]

In contrast to the usual semi-infinite diffusion condition we present a detailed mathematical analysis of grain boundary diffusion through a thin membrane of finite thickness in this paper. The membrane is bounded by two permeable and parallel surface planes. The particles diffuse into the sample through one surface and diffuse out of the sample through the opposite surface. In each case a constant flux is maintained through these surfaces during the entire diffusion process. Fick's second law of diffusion will be solved for a single isolated grain boundary as well as for multiple ones (i.e. an array of equidistant parallel grain boundaries). The exact mathematical solutions will be illustrated by several concentration profile plots for various times and appropriately chosen parameters. Additionally, we will calculate average concentration depth profiles in order to compare our calculations with experimental data.

2. BOUNDARY CONDITIONS

Diffusion processes arising in polycrystalline solids are usually described by Fick's second law for bulk diffusion

$$\frac{\partial c}{\partial t} = D\left(\frac{\partial^2 c}{\partial x^2} + \frac{\partial^2 c}{\partial y^2}\right) \tag{1}$$

and grain boundary diffusion

$$\frac{\partial c'}{\partial t} = D'\left(\frac{\partial^2 c'}{\partial x^2} + \frac{\partial^2 c'}{\partial y^2}\right) \tag{2}$$

where c, D, c' and D' denote the concentration of the diffusing species inside the grains (bulk), the bulk diffusion coefficient, the diffusant concentration within the grain boundaries and the grain boundary diffusion coefficient, respectively. The other symbols have their usual meanings. The boundary conditions concerning the geometric situation of the two models proposed here are summarised first; the boundary conditions concerning the transport processes between the grain boundaries and grains will be introduced separately.

The membrane is bounded by two parallel and permeable surface planes, a distance h apart. The diffusing particles flow into the grains of the solid through one surface ($y = 0$) and at the opposite surface ($y = h$) the mobile particles diffuse out of the grains. A constant flux J through these surfaces is maintained during the entire diffusion process. Moreover, the concentration of the mobile species in the layer at $y = h/2$ (parallel to the surface planes) may not vary during the diffusion process due to the symmetric geometry. Therefore, one arrives at the following boundary conditions

$$\frac{\partial c}{\partial y} = -\frac{J}{D}; \; y = 0, \, h \tag{3}$$

and

$$c - c_0 = 0; \; y = \frac{h}{2} \tag{4}$$

with c_0 being the equilibrium concentration of the diffusing particles at the initial time $t = 0$. It is worth mentioning that these conditions can be found in symmetric electrochemical cells consisting of powder compacts of mixed ionic–electronic conductors arranged between ionic conducting electrodes and probes, respectively. In this case the chemical diffusion in these mixed conducting compounds can be studied by means of galvanostatic polarisation/depolarisation experiments.[17–19]

2.1 *Isolated Grain Boundary*

In accordance with Fisher[3] as well as Gilmer and Farrell[7] the following model of an isolated grain boundary is proposed. The isolated grain boundary is imbedded perpendicularly to the surfaces of the membrane, as shown in Fig. 1. This grain boundary is assumed to be a highly diffusive and isotropic slab of a uniform thickness $2a$. The additional boundary condition arising from transport processes between the grain boundary and the adjacent grains reads[3,4]

$$D' \frac{\partial^2 c}{\partial x^2} - \frac{D}{a}\frac{\partial c}{\partial x} - \left(\frac{D'}{D} - 1\right)\frac{\partial c}{\partial t} = 0; \; x = a \tag{5}$$

assuming that the grain boundary thickness $2a$ is so thin that the concentration variation across its width (i.e. in the x-direction) is parabolic and the concentration of the diffusive species may be continuous at the grain boundary/crystal interfaces.

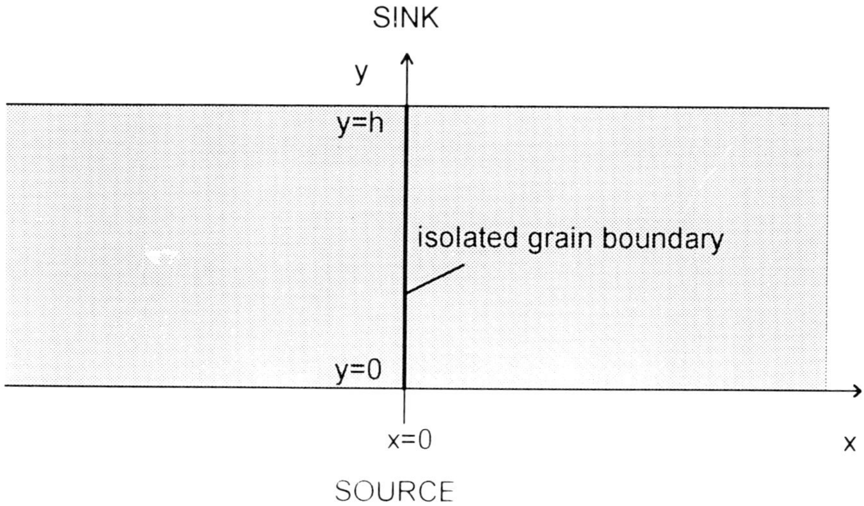

Figure 1 Schematic view of the isolated grain boundary model. One surface of the sample is in contact with the source of mobile species diffusing into the grains with a constant flux J, the opposite surface is in contact with the sink of the mobile species diffusing out of the grains with a constant flux J.

2.2 *Multiple Grain Boundaries*

Based on considerations performed by Gilmer and Farrell[8] the model of multiple grain boundaries is shown in Fig. 2. The grain boundaries are isotropic slabs imbedded perpendicularly to the surface planes and the distance between two neighbouring and parallel grain boundaries is L. The width $2a$ of these slabs is smaller by several orders of magnitude than the grain dimensions L and h, respectively. Assuming continuity and conservation of the diffusing species at the grain boundary/crystal interface, the boundary condition reads[8]

$$D' \frac{\partial^2 c}{\partial y^2} \mp \frac{D}{a} \frac{\partial c}{\partial x} = \frac{\partial c}{\partial t} \; ; \; x = \pm \frac{1}{2} L. \tag{6}$$

The concentration in the grain boundary is approximated by an expression with a parabolic dependence on the x-coordinate. Furthermore, the symmetric arrangement of the grain boundaries at $x = \pm L/2$ yields an additional boundary condition

$$\frac{\partial c}{\partial x} = 0; \; x = 0. \tag{7}$$

It must be emphasised that segregation of the diffusing species at the grain boundary/crystal interfaces has been neglected by the boundary conditions given above. This case may be described by $c' = sc$, with s being the segregation coefficient. As segregation of the mobile species generally complicates the grain

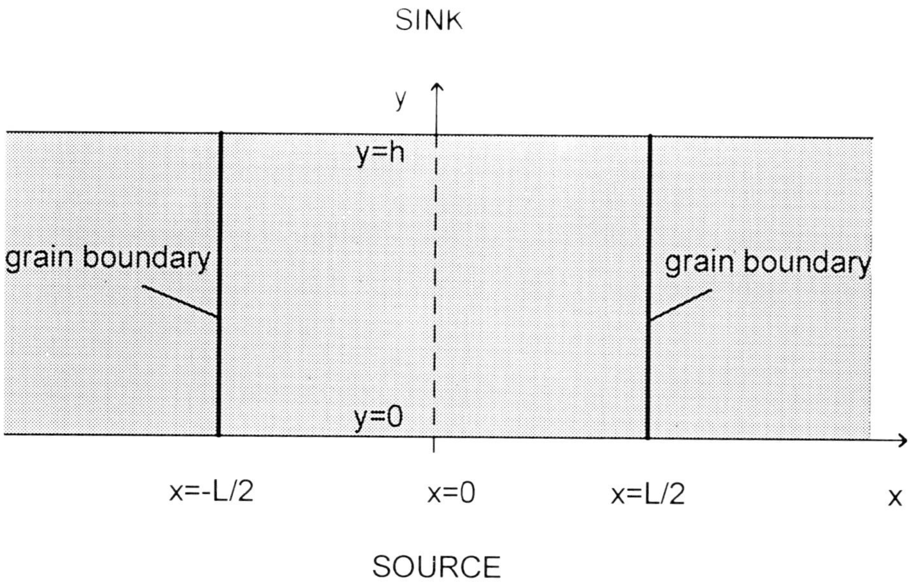

Figure 2 Schematic view of the multiple grain boundary model. One surface of the sample is in contact with the source of mobile species diffusing into the grains with a constant flux J, the opposite surface is in contact with the sink of the mobile species diffusing out of the grains with a constant flux J.

boundary diffusion process but does not influence the mathematical solution of the diffusion eq. (1), we need not consider this effect here.

3. MATHEMATICAL SOLUTION

Solving Fick's second law (eq. (1)) requires two different mathematical methods for the two different grain boundary models proposed above. In the case of an isolated grain boundary (Fig. 1) eq. (1) can be solved by applying Fourier–Laplace transforms whereas the analytical solution of eq. (1) for multiple grain boundaries (Fig. 2) is obtained by Fourier analysis techniques. A detailed description of the mathematical procedures for both cases is beyond the scope of this article. Thus, the mathematical solutions will be summarised in brief below whereas details can be found in Preis and Sitte.[20]

3.1 Solution for Isolated Grain Boundaries

According to Gilmer and Farrell[7] an adequate Fourier–Laplace transform taking into account the symmetry of the model shown in Fig. 1 is defined as

$$\Psi_n = \int_0^{\infty} e^{-\lambda t} \int_0^{h} \cos(\mu_n y)(c - c_0)\, dy\, dt. \tag{8}$$

The inverse transform reads

$$c - c_0 = \frac{2}{h} \sum_{n=1}^{\infty} \cos(\mu_n y) \frac{1}{2\pi i} \int_{-i\infty}^{+i\infty} e^{\lambda t} \Psi_n \, d\lambda \tag{9}$$

where μ_n is given by $(2n-1)\pi/h$ due to the boundary condition (4). The application of the transform (8) to eq. (1) as well as the boundary conditions (3) and (5) yields an ordinary differential equation system, the solution of which is given by

$$\Psi_n = \frac{2J}{\lambda D(\mu_n^2 + \lambda/D)} + \frac{2J(D'/D - 1)\exp[-\sqrt{\mu_n^2 + \lambda/D}(x-a)]}{D(\mu_n^2 + \lambda/D)\left(D' \mu_n^2 + \frac{D}{a}\sqrt{\mu_n^2 + \lambda/D} + \lambda \right)} \tag{10}$$

Introducing the following dimensionless parameters

$$\xi = \frac{x-a}{\sqrt{Dt}}, \ \eta = \frac{y}{\sqrt{Dt}}, \ \Delta = \frac{D'}{D}, \ \beta = \frac{a(\Delta - 1)}{\sqrt{Dt}}, \ \bar{\lambda} = \lambda t, \ \tilde{\mu}_n = \mu_n \sqrt{Dt} \tag{11}$$

and assuming $\Delta \to \infty$, while β remains finite, the inverse transform of Ψ_n can be written as

$$\begin{aligned} c - c_0 = {} & \frac{Jh}{D}\left(\frac{1}{2} - \frac{y}{h}\right) + \frac{4J}{hD} \sum_{n=1}^{\infty} \frac{Dt}{\tilde{\mu}_n^2} \cos(\tilde{\mu}_n \eta) \exp(-\tilde{\mu}_n^2) \\ & \times \left[\operatorname{erfc}\left(\frac{\xi}{2}\right) - \exp\left(-\frac{\xi^2}{4}\right) \exp\left(\frac{\xi}{2} + \tilde{\mu}_n^2 \beta\right)^2 \operatorname{erfc}\left(\frac{\xi}{2} + \tilde{\mu}_n^2 \beta\right) - 1 \right] \end{aligned} \tag{12}$$

by performing fundamental substitutions and transformations introduced by Whipple[4] and Gilmer and Farrell[7] (details can be found in ref. 20).

The variation of the average concentration in a layer parallel to the surface plane $\overline{c - c_0}$ with time t and position y is obtained by integrating eq. (12) from 0 to $L/2$ and multiplying by $2/L$, where L denotes the average grain size in x-direction, resulting in

$$\begin{aligned} \overline{c - c_0} = {} & \frac{Jh}{D}\left(\frac{1}{2} - \frac{y}{h}\right) + \frac{4J}{hD\Lambda} \sum_{n=1}^{\infty} \frac{Dt}{\tilde{\mu}_n^2} \cos(\tilde{\mu}_n \eta) \exp(-\tilde{\mu}_n^2) \left\{ \frac{2}{\sqrt{\pi}} \right. \\ & \times \left[1 - \exp\left(-\frac{\Lambda^2}{4}\right) \right] + \Lambda \operatorname{erfc}\left(\frac{\Lambda}{2}\right) - \frac{1}{\tilde{\mu}_n^2 \beta} \left[\exp\left(-\frac{\Lambda^2}{4}\right) \right. \\ & \left. \left. \exp\left(\frac{\Lambda}{2} + \tilde{\mu}_n^2 \beta\right)^2 \operatorname{erfc}\left(\frac{\Lambda}{2} + \tilde{\mu}_n^2 \beta\right) - \exp(\tilde{\mu}_n^4 \beta^2) \operatorname{erfc}(\tilde{\mu}_n^2 \beta) + \operatorname{erf}\left(\frac{\Lambda}{2}\right) \right] - \Lambda \right\} \end{aligned} \tag{13}$$

where Λ is defined as

$$\Lambda = \frac{L}{2\sqrt{Dt}} \tag{14}$$

Equation (12) shows the variation of the diffusant concentration as a function of time and position (x- and y-coordinate) for the case of an isolated grain boundary and eq. (13) gives the variation of the average concentration of diffusing particles with time and position (y-coordinate).

3.2 Solution for Multiple Grain Boundaries

Due to the symmetry of the model shown in Fig. 2 the following solution function $F(x, y, t)$ of eq. (1) can be found

$$F(x,y,t) = \sum_{n=1}^{\infty}\sum_{m=0}^{\infty} A_{mn}\cos(\alpha_{mn}x)\cos(\beta_n y)\exp[-Dt(\alpha_{mn}^2+\beta_n^2)]. \tag{15}$$

The parameter β_n is given by $(2n-1)\,\pi/h$, which follows from eq. (4). According to the boundary conditions (3) and (4) the general solution of eq. (1) for multiple grain boundaries can be written as

$$c-c_0 = \frac{Jh}{D}\left(\frac{1}{2}-\frac{y}{h}\right) - \sum_{n=1}^{\infty}\sum_{m=0}^{\infty} A_{mn}\cos(\alpha_{mn}x)\cos(\beta_n y)\exp[-Dt(\alpha_{mn}^2+\beta_n^2)]. \tag{16}$$

The coefficients A_{mn} are obtained by applying a Fourier analysis technique introduced by Gilmer and Farrell[8] (details can also be found in ref. 20), leading to

$$A_{mn} = \frac{16Jh}{L}\left[\frac{\sin(\alpha_{mn}L/2}{\alpha_{mn}} + a\cos(\alpha_{mn}L/2)\right] \Big/ \left\{\left[1+\frac{\sin(\alpha_{mn}L)}{\alpha_{mn}L}\right.\right.$$
$$\left.\left. + \frac{4a}{L}\cos^2(\alpha_{mn}L/2)\right] D(2n-1)^2\pi^2\right\}. \tag{17}$$

Inserting eq. (16) into the boundary condition (6) results in

$$(\Delta-1)\beta_n^2 - \alpha_{mn}^2 = \frac{\alpha_{mn}}{a}\tan(\alpha_{mn}L/2) \tag{18}$$

which represents a transcendental expression for the parameter α_{mn}.

The average concentration of the mobile species is obtained by the same procedure as already described in the case of an isolated grain boundary, i.e. integrating eq. (16) from $x = 0$ to $x = L/2$ and multiplying by $2/L$

$$\overline{c-c_0} = \frac{Jh}{D}\left(\frac{1}{2}-\frac{y}{h}\right) - \frac{1}{2}\sum_{m,n}\frac{A_{mn}}{\alpha_{mn}}\sin(\alpha_{mn}L/2)\cos\left[\frac{(2n-1)\,\pi y}{h}\right]$$
$$\times\exp\left[-Dt\left(\alpha_{mn}^2+\frac{(2n-1)^2\pi^2}{h^2}\right)\right]. \tag{19}$$

The variation of the diffusant concentration with time and position (x, y) is

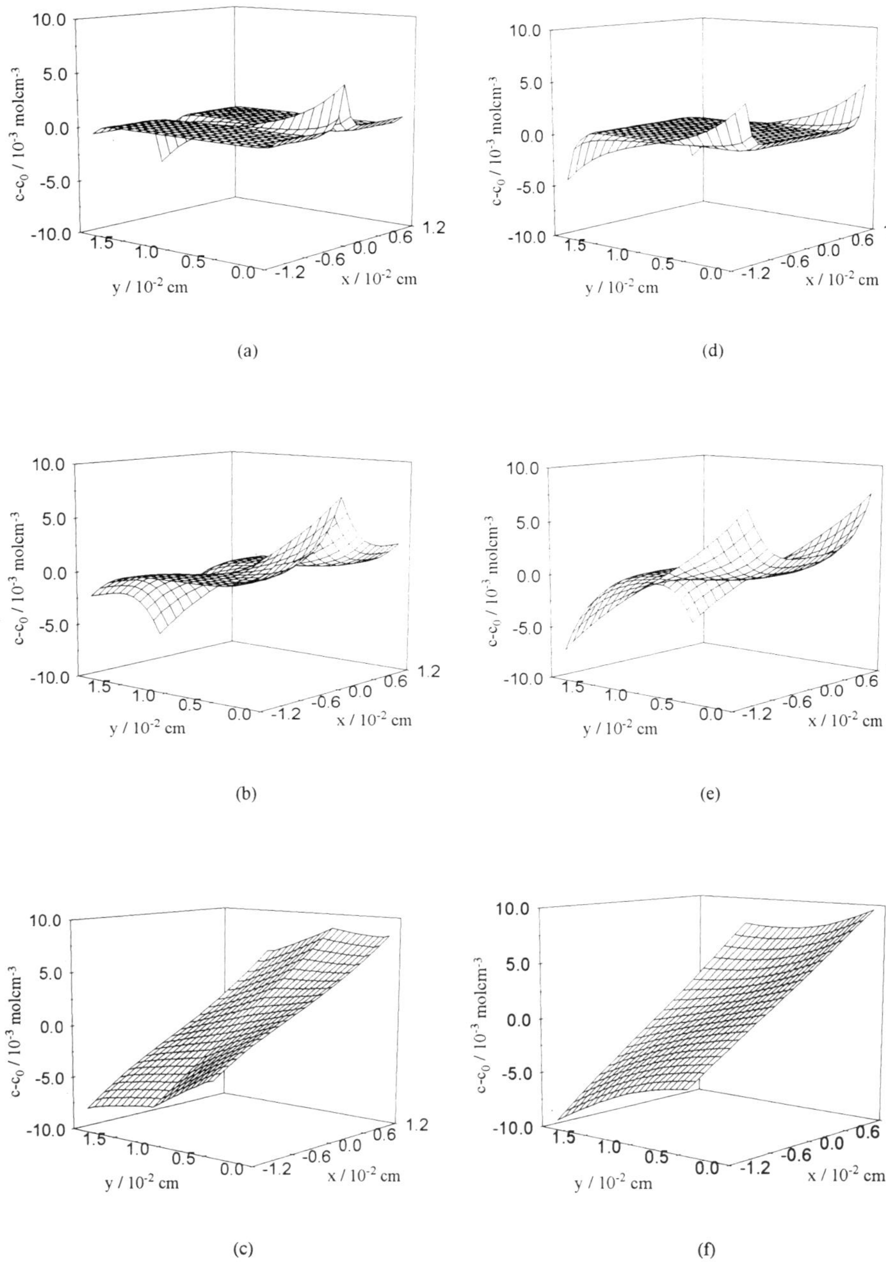

Figure 3 Concentration profile plots at 50 s (a, d), 500 s (b, e) and 5000 s (c, f). Left-hand figures represent the case of an isolated grain boundary (a–c), right-hand figures show profiles for multiple grain boundaries (d–f). $L = 0.02$ cm; $J = 10^{-8}$ mol cm^{-2} s^{-1}; $a = 2.5 \cdot 10^{-8}$ cm; $h = 0.02$ cm; $D = 10^{-8}$ cm^2 s^{-1}; $\Delta = 10^6$.

obtained from eq. (16), where the parameters α_{mn} are calculated numerically from the transcendental eq. (18) and the coefficients A_{mn} are obtained from eq. (17). The average concentration profiles can be calculated by application of eq. (19) in combination with eqs. (17) and (18). Again, details about the necessary numerical operations can be found in ref. 21.

4. RESULTS AND DISCUSSION

All concentration profile plots have been calculated for both models by choosing the following values for the parameters introduced in the section above. The flux J is assumed to be 10^{-8} mol cm^{-2} s^{-1}, the bulk diffusion coefficient D is 10^{-8} cm^2 s^{-1}, the grain boundary width $2a$ is 5×10^{-8} cm and the length h is 0.02 cm.

In Fig. 3 several concentration profiles for various times (50, 500 and 5000 s) are shown. The left-hand column of Fig. 3 (a–c) describes an isolated grain boundary whereas the right-hand column (d–f) considers the case of multiple grain boundaries. The grain boundary/bulk diffusion coefficient ratio Δ amounts to 10^6 and the grain boundary distance L is 0.02 cm. The profiles shown in Fig. 3 show a strong influence of the grain boundaries on the overall diffusion through a polycrystalline solid. Particularly at short times (50 s) the transport in the region around the grain boundary is considerably enhanced, as the grain boundary diffusion coefficient D' is chosen to be higher by several orders of magnitude than the bulk diffusion coefficient D, see Figs 3(a) and 3(d). After sufficiently long diffusion anneal times (5000 s) the diffusion through the bulk of the sample has proceeded so far that the enhancement of the transport due to grain boundary diffusion has been reduced resulting in an almost vanishing concentration gradient along the x-coordinate ($\partial c/\partial x \approx 0$), see also Figs 3(c) and 3(f).

The transport along an isolated grain boundary can be referred to Harrison's type B-kinetics,[1,10,11,22] i.e. the overlap of the diffusion fields of neighbouring grain boundaries is completely neglected. This assumption is fulfilled if the condition

$$Dt \ll L^2 \tag{20}$$

is satisfied. On the other side, the concept of multiple grain boundaries is a physically more reliable model of a polycrystalline membrane, since the interaction of transport processes arising at adjacent grain boundaries is taken into account.

At two different times (50 and 500 s) average concentration profile plots for an isolated grain boundary are compared with those for multiple grain boundaries in Figs 4–6. In Fig. 4 the variation of the average diffusant concentration $\overline{c - c_0}$ with position y is presented, assuming $\Delta = D'/D = 10^6$ and $L = 0.02$ cm. The average concentration profile for isolated grain boundaries (dashed line) differs somewhat from the profile for the multiple case (solid line) at 500 s whereas at 50 s, i.e. short diffusion times, both curves coincide fairly well. A perfect coincidence is found for both times (50 and 500 s) in Fig. 5, where the relevant parameters are $\Delta = 10^6$ and $L = 0.1$ cm, i.e. the grain boundary distance L (average grain size) has been

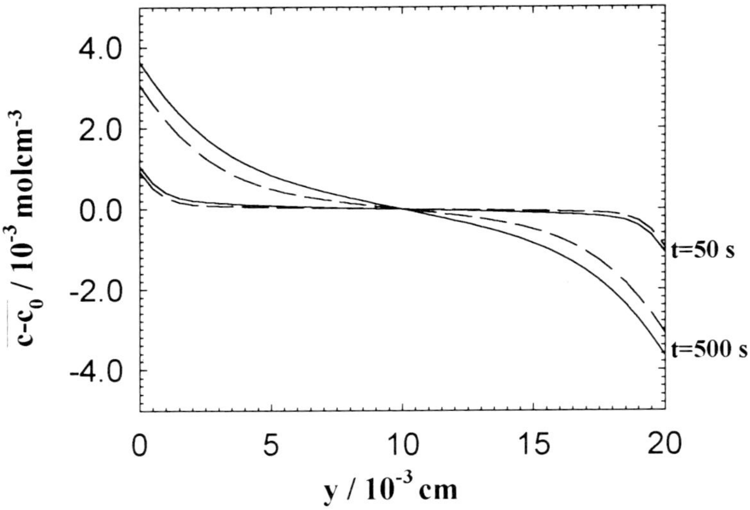

Figure 4 Average concentration profile plots for an isolated (dashed lines) and multiple (solid lines) grain boundaries at 50 s and 500 s. $L = 0.02$ cm, $\Delta = 10^6$; $J = 10^{-8}$ mol cm^{-2} s^{-1}; $a = 2.5 \cdot 10^{-8}$ cm; $h = 0.02$ cm; $D = 10^{-8}$ cm^2 s^{-1}.

increased. In addition, the concentration profile plots concur very well, if the following values are chosen, viz. $\Delta = 10^5$ and $L = 0.02$ cm (the grain boundary diffusion coefficient D' has been decreased while the bulk diffusion coefficient D remains constant).

As a consequence the average concentration profile plots concur very well when the diffusion time is quite short and/or the grain boundary distance (average grain size) is high. These results confirm the validity of condition (20) which confines the applicability of the isolated grain boundary model. The good coincidence between the curves corresponding to the decreased value of the parameter Δ can be interpreted in terms of a reduced influence of the grain boundaries on the overall diffusion process through the membrane. Hence, the difference between the particular models is fairly unimportant, as the significance of grain boundaries for the transport process will generally be reduced owing to the decrease of Δ.

5. CONCLUSION

Analytical solutions for grain boundary diffusion through membranes containing isolated as well as multiple grain boundaries have been derived by application of Fourier–Laplace transforms and Fourier analysis techniques. Both solutions have been used to calculate concentration profiles. The average concentration depth profile plots obtained for an isolated grain boundary have been compared with profiles received for multiple grain boundary diffusion, leading to the result that the condition $Dt \ll L^2$ must be satisfied in order to guarantee a satisfactory coincidence

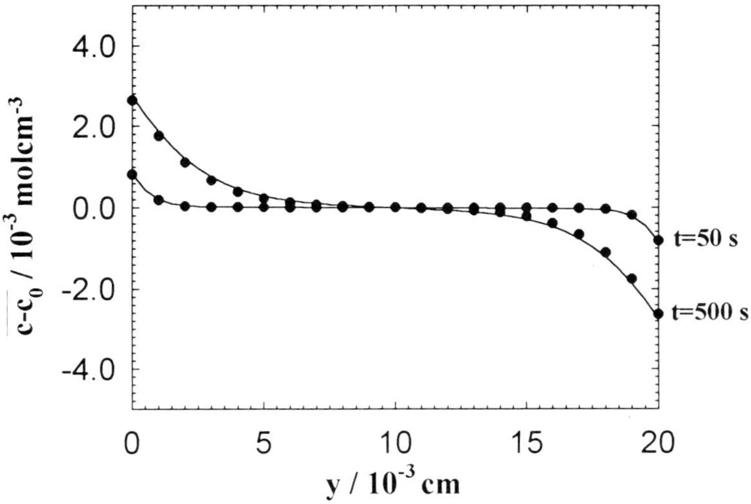

Figure 5 Average concentration profile plots for isolated (filled circles) and multiple (solid lines) grain boundaries at 50 s and 500 s. $L = 0.1$ cm; $\Delta = 10^6$; $J = 10^{-8}$ mol cm^{-2} s^{-1}; $a = 2.5 \cdot 10^{-8}$ cm; $h = 0.02$ cm; $D = 10^{-8}$ cm^2 s^{-1}.

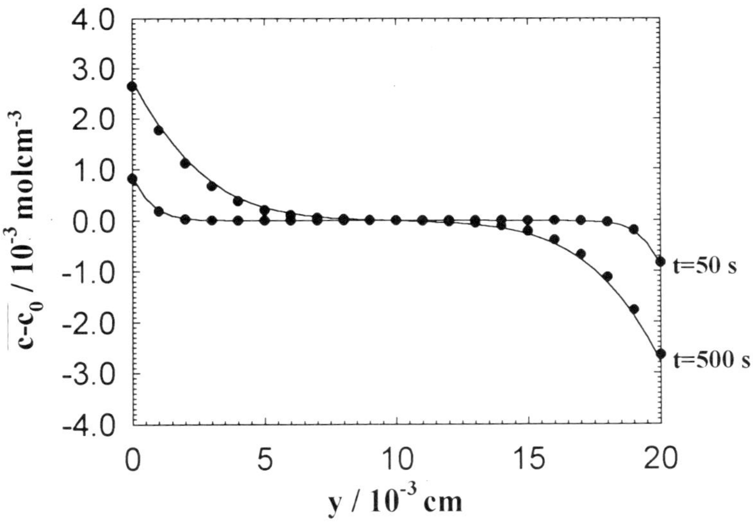

Figure 6 Average concentration profile plots for isolated (filled circles) and multiple (solid lines) grain boundaries at 50 s and 500 s. $L = 0.02$ cm; $\Delta = 10^5$; $J = 10^{-8}$ mol cm^{-2} s^{-1}; $a = 2.5 \cdot 10^{-8}$ cm; $h = 0.02$ cm; $D = 10^{-8}$ cm^2 s^{-1}.

between the average concentration profiles of both models. The boundary conditions proposed in the present investigation (constant fluxes through both surfaces during the entire diffusion process) may be found in the case of ionic or mixed ionic–electronic conducting ceramic membranes used in solid oxide fuel cells or ceramic oxygen sensors or generators. Additionally, the boundary conditions apply to galvanostatic polarisation experiments on symmetric electrochemical cells consisting of a mixed ionic–electronic conducting sample as performed in the past by us. Neglecting any polarisation processes caused by electrode reactions, the chemical diffusion coefficient of the mixed conductor can in this case be determined from the time behaviour of the cell voltage. As this time behaviour is given by the variation of the diffusant concentration with time at the voltage probes, our solution of grain boundary diffusion can supply appropriate fitting procedures for the determination of grain boundary as well as bulk diffusion coefficients of mixed conductors by means of such electrochemical polarisation/depolarisation experiments.

REFERENCES

1. I. Kaur and W. Gust: *Fundamentals of Grain and Interphase Boundary Diffusion*, 1988, Ziegler Press, Stuttgart.
2. B. C. H. Steele: *Materials Sci. Eng.*, 1992, **B13**, 79.
3. J. C. Fisher: *J. Appl. Phys.*, 1951, **22**, 74.
4. R. T. P. Whipple: *Phil. Mag.*, 1954, **45**, 1225.
5. T. Suzuoka: *Trans. Japan Inst. Metals*, 1961, **2**, 25.
6. A. D. Le Claire: *Brit. J. Appl. Phys.*, 1963, **14**, 351.
7. G. H. Gilmer and H. H. Farrell: *J. Appl. Phys.*, 1976, **47**, 3792.
8. G. H. Gilmer and H. H. Farrell: *J. Appl. Phys.*, 1976, **47**, 4373.
9. J. C. M. Hwang and R. W. Balluffi: *J. Appl. Phys.*, 1979, **50**, 1339.
10. D. Gupta, D. R. Campbell and P. S. Ho: *Thin Films Interdiffusion and Reactions*, J. M. Poate, K. N. Tu and J. W. Mayer (eds), 1978, p. 161, Wiley, New York.
11. N. L. Peterson: *Internl Metals Rev.*, 1983, **28**, 65.
12. X. Gui, S. K. Dew, M. J. Brett and D. de Cogan: *J. Appl. Phys.*, 1993, **74**, 7173.
13. J. Yao and J. R. Cahoon: *Acta Metall. Mater.*, 1991, **39**, 111.
14. B. C. H. Steele, J. A. Kilner, P. Dennis and D. S. Tannhauser: *Brit. Ceram. Proc.*, 1984, **34**, 53.
15. R. J. Chater, S. Carter, J. A. Kilner and B. C. H. Steele: *Solid State Ionics*, 1992, **53–56**, 859.
16. D. Grientschnig and W. Sitte: *J. Phys. Chem. Solids*, 1991, **52**, 805.
17. W. Sitte: *Solid State Ionics*, 1993, **59**, 117.
18. I. Yokota: *J. Phys. Soc. Japan*, 1961, **16**, 2213.
19. I. Rom and W. Sitte: *Solid State Ionics*, 1994, **70–71**, 147.
20. W. Preis and W. Sitte: 'Grain boundary diffusion through thin films. Application to permeable surfaces', *J. Appl. Phys.*,in press.
21. L. G. Harrison: *Trans. Faraday Soc.*, 1961, **57**, 1191.

Concentration and Transport of Protons and Oxygen Defects in Oxides

TRULS NORBY and YNGVE LARRING
Centre for Materials Research, University of Oslo, Gaustadalleen 21, N-0371 Oslo, Norway

ABSTRACT
The paper discusses the concentration and mobility of protons relative to oxygen vacancies in oxides. The relative concentration of protons increases with increasing stability of the oxide. In particular, we have for rare earth sesquioxides R_2O_3 and perovskites $R'R''O_3$ found a good empirical correlation between the equilibrium constant for the reaction $H_2O(g) + V_O^{\cdot\cdot} = 2H_i^{\cdot} + O_O^x$ and a parameter such as molar density which expresses bond strength between oxygen and metal ions. It is thus possible to estimate equilibrium concentrations of protons and oxygen vacancies for acceptor substituted oxides in this group. Proton transport is discussed in the light of recent results in modelling, neutron scattering, isotope effects and concentration cell studies. Protons migrate by the free proton jump mechanism, and proton transfer is facilitated by thermal modes in the host oxygen lattice. This explains the activation energies similar to and generally somewhat lower than for oxygen vacancy transport, without the need for tunnelling or hydroxide transport as sometimes proposed in the literature.

1. INTRODUCTION

As suggested by Stotz and Wagner,[1] oxides equilibrate with wet atmospheres at high temperatures under dissolution of protons. The protons bond to oxygen ions, forming substitutional hydroxyl defects, $OH_O^{\cdot}$. However, we may for simplicity consider the oxygen ion a part of the perfect lattice and denote the defect an interstitial proton, $H_i^{\cdot}$.

The positive charge of the protons affects the electroneutrality of the material, and defect-dependent properties such as conductivity, diffusivity, mechanical strength and corrosion resistance may become functions of the concentration of protons and, in turn, of parameters such as $p(H_2O)$.[2]

Particularly interesting and important is the competition between oxygen vacancies and protons. For instance, acceptor doping will, depending on $p(H_2O)$, T, and the doping level, be compensated mainly by oxygen vacancies or protons in many oxides. We have for some time been looking for a correlation between this balance and other materials properties. A correlation to the molar (packing) density for many (mainly rare earth) oxides of general formula ABO_3 (including A_2O_3) has been found and will be presented, enabling estimation of proton (and oxygen vacancy) concentration for a given oxide, doping, $p(H_2O)$ and temperature. This is important for selecting candidates for solid electrolytes and for correct interpretation in studies of non-stoichiometry and mass transport.

Oxides with low proton diffusivities may be used as hydrogen barriers,[3] while oxides with fast proton transport may be used as proton-conducting electrolytes.[4]

The ionic conduction in many oxides, earlier believed to reflect oxygen ion transport, can now be assigned to protons.

EMF-method results and H/D isotope effect studies indicated early that protons in oxides are transported as protons diffusionally jumping between stationary oxygen ions. However, the low activation energies of protonic transport have nourished models employing tunnelling of protons as well as hydroxide ion transport. We will briefly review recent advances in theory and experiment that enable us to understand the observed properties in terms of the free proton diffusion model.

2. THE EQUILIBRIUM BETWEEN PROTONS AND OXYGEN VACANCIES

2.1 *Defect Chemistry*

The formation of oxygen vacancies can in terms of the Kröger–Vink notation be written as

$$O_O^x = V_O^{\cdot\cdot} + 2e' + 1/2O_2(g) \qquad K_1 = [V_O^{\cdot\cdot}][e']^2 p(O_2)^{1/2} \tag{1}$$

where the equilibrium constant has been written assuming low defect concentrations, i.e. $[O_O^x]$ is constant. In the presence of water vapour, protons may be incorporated:[1]

$$H_2O(g) = 2H_i^{\cdot} + 2e' + 1/2O_2(g) \qquad K_2 = [H_i^{\cdot}]^2[e']^2 p(O_2)^{1/2} p(H_2O)^{-1} \tag{2}$$

For our purpose it is practical to combine eqs (1) and (2) and obtain[1]

$$V_O^{\cdot\cdot} + H_2O(g) = 2H_i^{\cdot} + O_O^x \qquad K_3 = K_2/K_1 = [H_i^{\cdot}]^2[V_O^{\cdot\cdot}]^{-1} p(H_2O)^{-1} \tag{3}$$

Finding the value of K_1 or the enthalpy change dH_1 involved is not simple, and these values are in general unknown, although Kofstad[5] suggested empirical relations between dH_1 and the enthalpy of formation and atomisation of the oxide. Similarly, K_2 and dH_2 are generally unknown. dH_2 contains well-known enthalpies of the gas phase reactions involved, and Norby and Kofstad[6] suggested that the additional dissolution of hydrogen atoms and their ionisation to protons and electrons require little further energy, and with modest variations between oxides. Experiments and calculations indicate that dissolution and ionisation of $2H$ in Er_2O_3 contributes around −90 kJ/mol (−45 kJ/mol H),[7] and similarly that only a few tens of kJ/mol are involved in the case of Y_2O_3.[8] Values for other oxides are not available.

K_3 and dH_3 determine the equilibrium between oxygen vacancies and protons and can be relatively readily determined by measurements on acceptor doped oxides: Consider an oxide doped with a lower-valent cation Ml, forming the defect Ml'_M, compensated by protons and oxygen vacancies. The electroneutrality condition can then be approximated with

$$2[V_O^{\cdot\cdot}] + [H_i^{\cdot}] = [Ml'_M] = \text{constant} \tag{4}$$

Figures 1 and 2 show how defect concentrations behave under oxygen vacancy and proton dominance as a function of, respectively, $p(H_2O)$ and temperature. Based on experimental evidence, Fig. 2 is drawn assuming that the enthalpy change of eq. (3), dH_3, is negative. We see that protons dominate over oxygen vacancies at relatively high $p(H_2O)$ and low temperature. (There is a change in defect dominance also as a function of $[Ml'_M]$: Protons dominate at low acceptor levels and oxygen vacancies at high.[7])

Due to the change in defect structure, the value of K_3 and dH_3 can be determined by measuring defect-dependent properties vs these parameters. For instance, at given T and $[Ml'_M]$, the changeover takes place at $p(H_2O)_t$, and this can be taken as the point where $2[V_O^{\cdot\cdot}] = [H_i^{\cdot}] = [Ml'_M]/2$, so that

$$p(H_2O)_t = [Ml'_M]/K_3 \tag{5}$$

Thus, K_3 and dH_3 can be determined from the bend point $p(H_2O)_t$ in electrical conductivity vs $p(H_2O)$ at different temperatures, provided the acceptor concentration $[Ml'_M]$ is known. ($p(O_2)$, on the other hand, has no influence on the relative concentrations of oxygen vacancies and protons; measurements vs $p(O_2)$ at constant $p(H_2O)$ give no information about the role of protons or of K_3.)

dH_3 can be found also by subtracting the activation enthalpies for a partial conductivity in the region of proton dominance from that in the region of oxygen vacancy dominance (the activation energy for the mobility cancels). By doing the subtraction for oxygen vacancy conduction, dH_3 is obtained, while for protons and electron holes one gets $dH_3/2$.[9]

The parameters for K_3 and other equilibrium constants as well as defect mobilities may also be found by computer fitting of the defect model to a sufficiently large set of total or partial conductivities under different conditions.[10]

Of course, K_3 can be found also if the absolute concentration of oxygen vacancies and/or protons can be found by some suitable method and inserted into eqs (3) and (4).

2.2 K_3 *in Different Oxides*

Figure 3 shows estimates of K_3 vs $1/T$ for a number of oxides, based on various experimental studies. K_3 is here expressed in terms of reference activities of 1 mol defects/mol oxide and 1 atm H_2O. The values in Figs 3 and 4 are extracted from ref. 9 (Sm_2O_3), ref. 10 ($LaErO_3$), ref. 11 (Er_2O_3, Dy_2O_3, Gd_2O_3), ref. 12 (Y_2O_3), ref. 13 (La_2O_3, Nd_2O_3), ref. 14 ($SrCeO_3$), ref. 15 ($BaCeO_3$), and ref. 16 (Al_2O_3). K_3 decreases with increasing temperature for all oxides in Fig. 3, i.e. dH_3 is negative. The highest K_3 is found for Er_2O_3, with $dH_3 = -210$ kJ/mol. $SrTiO_3$, on the other

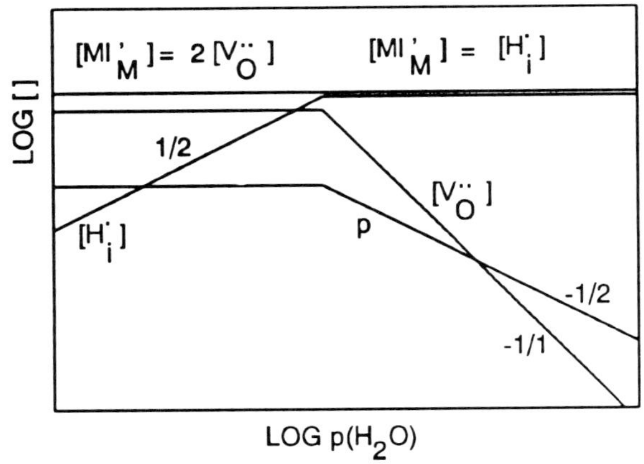

Figure 1 Schematic diagram of defect concentrations vs $p(H_2O)$.

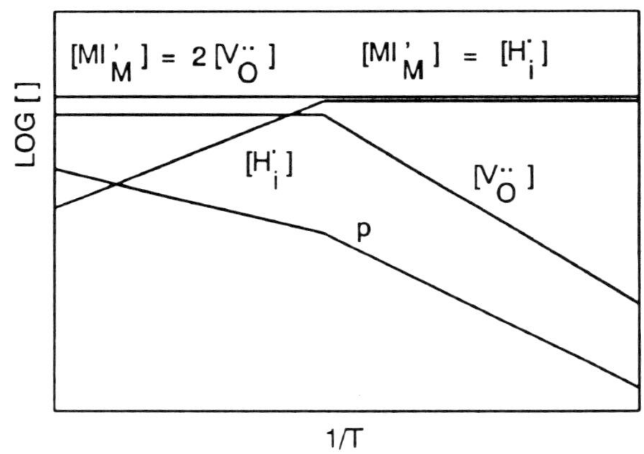

Figure 2 Schematic diagram of defect concentrations vs $1/T$.

hand, has a value of dH_3 of only around -30 kJ/mol,[17] giving low relative proton concentrations, and no reported values of K_3.

2.3 Relations Between K_3 *and* dH_3 *and Other Materials Properties*

With so many results on K_3 and dH_3 it is of interest to ask whether they correlate with other properties of the oxides. Correlation of dH_3 with estimates of dH_1 and dH_2 give reasonable results only in selected cases, as mentioned earlier. The general failure probably arises from our inability to predict dH_1 because of inaccuracies in the empirical relation or in the available thermodynamic data involved. In the following we therefore try instead to relate K_3 and dH_3 directly with other more well-known materials properties which assumingly are related to the enthalpy dH_1 of formation of oxygen vacancies.

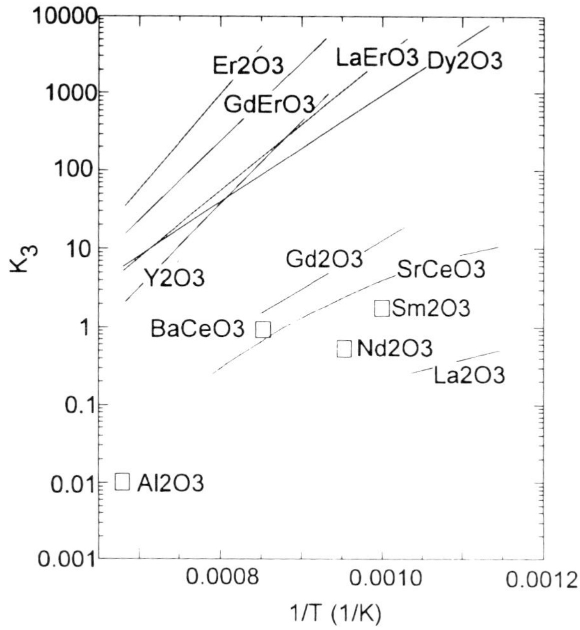

Figure 3 Estimates of K_3 vs $1/T$ for some acceptor-doped oxides (refs in the text).

The more stable the oxide, the harder it must be to form a vacancy. Thus, dH_1 and, in turn, dH_3 and K_3, may be coupled to the enthalpy of formation of the oxide. Figure 4 shows K_3 at 977°C (estimated from Fig. 3) vs enthalpy of formation (from ref. 18, except for $SrCeO_3$, ref. 19).

Despite the scatter, Fig. 4 suggests a certain coupling between K_3 and the enthalpy of formation of the oxide. The correlation is necessarily blurred by differences in the energies in the standard states of the metals.

The bond strength between metal and oxygen may be a more appropriate measure of the energy of formation of oxygen vacancies. Furthermore, the bond strengths also determine the packing density of the oxide. So within comparable structural groups the easily accessible and accurate molar densities may provide a good correlation with dH_1 and, in turn, dH_3. Larring and Norby[11] plotted dH_3 vs molar density for a number of rare earth binary and ternary oxides, see Fig. 5. The correlation is good, and linear regression gave:

$$dH_3(\text{eV}) = 9.4 \pm 0.7 - 510 \pm 35\rho \tag{6}$$

where ρ is the molar density in mol/cm^3. Furthermore, it was found that the pre-exponentials for the different oxides were in good agreement, such that we can generalise further:

$$K_3(\text{atm}^{-1}) = 2 * 10^{-6} \exp(-dH_3/\text{k}T) \tag{7}$$

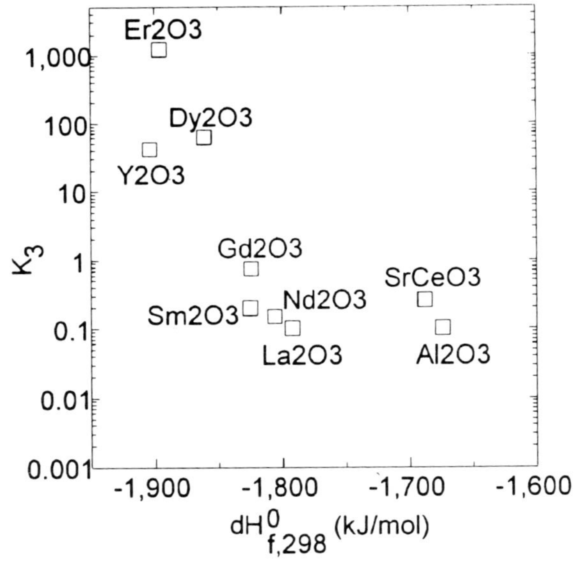

Figure 4 Estimates of K_3 at 977°C vs enthalpy of formation (refs in the text).

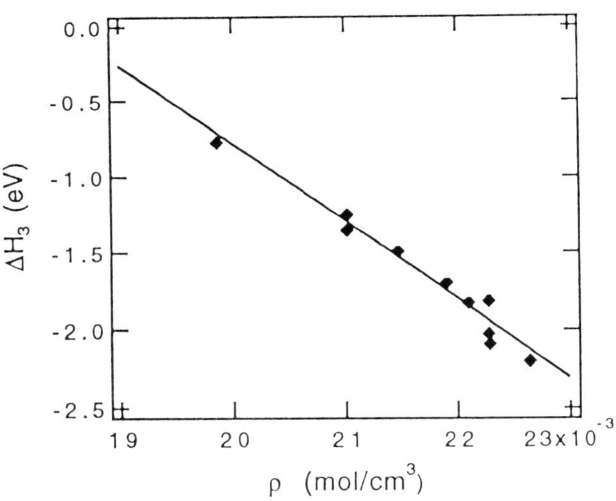

Figure 5 dH_3 vs molar density for a number of acceptor substituted rare earth binary and ternary oxides. Extracted from ref. 11.

where dH_3 is given by eq. (6). Perovskites containing other elements than the rare earths, like alkaline earth cerates and titanates, appear to follow the same trend as the rare earth oxides, but more work is needed to determine these correlations as well as correlations for other structural groups.

The possibility to estimate K_3 from the molar density will enable calculations to be made of proton and oxygen vacancy concentrations for oxides and conditions where these have not been measured. For this purpose we may combine eqs (3) and (4), solve with respect to the concentration of the defect of interest, and insert K_3 according to such expressions as eqs (6) and (7).

3. PROTONIC TRANSPORT

3.1 Mechanisms of Protonic Transport

In describing proton conductors one usually distinguishes between the free proton transport mechanism (Grotthuss type) and proton transport by vehicle mechanisms. In free proton transport, protons jump between stationary oxygen ions. Each jump is followed by a rotation around the oxygen ion to get in position for the next jump. The jump is normally considered the rate-limiting step; the rotation is easier.

In vehicle mechanisms, the proton moves as a passenger on a larger species like oxygen ions. We can thus envisage transport by hydroxyl, water, or hydronium ions when one, two, or three protons, respectively, travel with the oxygen ion. As discussed below, vehicle mechanisms can only account for dominant long-range transport processes in oxides when proceeding on interstitial lattice sites. Therefore, proton conducting perovskites can hardly be expected to be so by means of hydroxyl ion transport, as sometimes suggested in the literature.

In the following section we briefly review methods for experimentally separating protonic and native ionic transport and for identifying the transported protonic species.

3.2 Experimental Determination of Protonic Transport

The three isotopes of hydrogen offer several exchange and detection methods for diffusivity measurements. For instance, Al_2O_3, BeO, and Y_2O_3 were found to exhibit low diffusivities of recoil-injected tritium.[3] Hydrogen isotope diffusivities have been measured with high accuracy, and are in good agreement with a detailed diffusion model, for TiO_2, rutile.[20]

The dominant charge carrier can be identified by characterising the chemical output of a fuel cell or electrolyser when DC current is passed. This method has shown many perovskite oxides to exhibit oxygen ion transport at high temperatures and proton transport at more moderate temperatures.[21]

The concentration cell (EMF or OCV) method directly yields the transport number of protonic defects when used with a gradient in $p(H_2O)$ or $p(H_2)$ without a gradient in $p(O_2)$.[22] It can also identify the migrating protonic species. For instance, the sign of the voltage in a gradient of $p(H_2O)$ reverses for OH^- transport compared to H^+ transport. The method and theory behind it have been received with some scepticism, and the results, which in the literature always point at H^+ transport, have been neglected by many who are in favour of OH^- transport. However, the equations used have been confirmed by different derivations,[23,24] and the proof of H^+ transport obtained by this method cannot be neglected.

It may be noted that ^{18}O diffusivity measurements will detect both O^{2-} and OH^- transport. By comparison with the much higher total transport data in the case of $BaCeO_3$, this was used to eliminate the possibility of an OH^- transport mechanism for protons in this material.[25]

The H/D isotope effect in conductivity measurements is another useful check of free proton transport: The O–H bond oscillates at a frequency inversely proportional to the square root of its reduced mass, $m_r = 1 * 16/(1 + 16) = 0.94$. The O–D bond correspondingly has $m_r = 2 * 16/(2 + 16) = 1.78$. The ratio between the inverse, square root of the two is 1.37. This is the ratio between the attempt frequencies of H and D to break the bond to the oxygen ion, and the ratio should be reflected in the pre-exponentials of proton and deuteron conductivity. If OH^- was the migrating species, the isotope effect would be much lower; a factor of around 1.03. With few (and often doubtful) exceptions, the observed ratios are as expected, or higher, for free proton transport when the materials are expected to be proton conductors.

It is common to find that the activation energy for deuteron transport is 0.04–0.06 eV higher than for proton transport, and this further increases the isotope effect. This can be ascribed to the 0.055 eV difference in zero-point energy between the OH^- and OD^- oscillators.[20,26] Underlying minority (hole and oxygen vacancy) conductivities with higher activation energies will furthermore contribute relatively more to deuteron than proton conductivity, giving too small isotopic differences in pre-exponential factor and too large differences in activation energy (unless properly subtracted).[27]

Other techniques may also give useful information about protonic transport. NMR is well known in terms of short-range transport. Examples of new techniques being implemented are internal friction measurements[28] and quasielastic neutron scattering,[29] and particularly the latter seems to contribute valuable information on details in proton transport mechanisms in well-chosen materials.

3.3 OH^- Transport in Oxides

Most investigators assume that protons move as free H^+ ions, and those who have tested this experimentally always conclude the same. The doubts mostly come from investigators noting that protonic transport has activation energies following relatively closely those of oxygen transport. Furthermore, protonic mobility generally increases with increasing lattice parameters and O–O distances. Intuitively, these observations are expected for OH^- transport, but not for free proton transport. Nevertheless, we have experimental evidence for free proton transport from DC measurements, concentration cell studies, and isotope effects in a number of oxides.

Moreover, consider a close-packed structure, like a perovskite, under conditions where protonic conduction dominates. If a substitutional OH^- is located next to an oxygen vacancy, it can surely move by jumping into that vacancy. For the next step there are three possibilities: (1) the OH^- jumps back to the empty original position; (2) the proton jumps to another oxygen ion; a new OH^- jump can be made only when the proton arrives on an oxygen ion neighbouring a vacancy; (3) the vacancy moves away from the OH^-; the OH^- can only jump again when another vacancy arrives. Clearly, case 1 makes no contribution to long-range transport, case 2 is mainly free proton transport, and case 3 is mainly oxygen ion transport. We conclude that transport by substitutional OH^- (or other vehicles) cannot be a dominant transport mechanism.

Case 2 above can nevertheless provide proton-aided oxygen transport: the hydroxyl ion, because of its smaller charge, may have a smaller activation energy than a normal oxygen ion for jumping into a neighbouring vacancy. In a material with slow oxygen transport and faster proton transport, a relatively small number of protons may thus diffuse around and 'catalyse' oxygen transport by OH^- jumps.[2,30]

In oxides with interstitial oxygen sites, long range transport by interstitial OH^- can be envisaged, and in this case OH^- transport can be a dominating mechanism. However, not even in cubic rare earth sesquioxides, with ample interstitial oxygen sites, have either dominant interstitial hydroxyl defects or transport by OH^- species been detected.

3.4 Free Proton Transport

Based on the above discussion, we focus on the observed features of protonic transport in terms of the free proton transport mechanism. The activation energies of proton mobility range from around 0.5 eV and upwards. They are typically 2/3 of the activation energy for oxygen vacancy mobility, and proton diffusivity has, for instance, been found to be 2 orders of magnitude higher than oxygen diffusivity in $BaCeO_3$ at 650°C.[31] The activation energy decreases and proton mobility increases with increasing O–O distance, leading some authors through the years to suggest that proton jumps between oxygen ions cannot be rate limiting. However, a static lattice simulation of proton jumps in perovskites predicted the observed dependency on O–O distance, but with too high activation energies.[32]

In addition to suggestions of OH^- transport as discussed above, it has also been suggested occasionally that the low activation energies can be due to proton tunnelling. However, hydrogen tunnelling is commonly considered significant only around room temperature or below. Furthermore, deuterium tunnelling is even less likely, so tunnelling would have given anomalously high isotope effects rather than the close to classical effects seen in practice. The existence of high temperature tunnelling has been defended by some in light of the report of two orders of magnitude higher proton conductivity in $CsHSO_4$ relative to $CsDSO_4$.[33] However, this refers to very low conductivities and temperatures below 140°C, and could reflect sample differences.[34] Conductivity measurements on $CsHSO_4/CsDSO_4$ with *in situ* isotopic exchange have not confirmed the anomalous isotopic effect.[35]

Kreuer *et al.*[36] recently modelled proton transport in oxides and concluded that proton transfer between oxygen ions is facilitated by the thermal fluctuations of the two oxygen ions: when the oxygen ions approach each other, the energy barrier for proton transfer diminishes. In ABO_3 perovskites, this is most favourable for the two oxygen ions taking part in the O–B–O bending mode. For good proton conductors, such as the cerates, this mode is 'soft' enough to explain the low activation energy. Furthermore, this model points at 'softness' in the oxygen sublattice as the most important parameter for proton transport in oxides, favouring large polarisable cations and large O–O distances, exactly as observed in practice.

Recently, Hempelmann *et al.*[29] investigated proton diffusion in Yb-substituted $SrCeO_3$ using quasielastic neutron scattering (QENS) at high temperatures. Excellent agreement was found with conductivity data in terms of activation energy of around 0.6 eV. The results furthermore indicated that the activation energy for proton transfer between oxygen ions was even lower, and that the observed average activation energy was a result of trapping at oxygen ions next to the Yb acceptors.

4. CONCLUSIONS

The relative concentration of protons at the cost of oxygen vacancies increases with increasing thermodynamic stability of the oxide. In particular, the equilibrium between protons and oxygen vacancies seems well correlated to the packing (molar) density within comparable groups of oxides such as rare earth sesquioxides and perovskites. Such correlations enable estimation of concentrations of oxygen vacancies and protons in acceptor-doped oxides under all conditions, based simply on structural data. This will help in the selection of solid electrolyte candidates and in proper interpretation of measurements of transport properties and stoichiometry. Protonic transport takes place by the free proton jump mechanism. Hydroxyl ion transport cannot provide a dominant conductivity in a close-packed structure, but can accomplish minority oxygen transport in a proton conductor.

The free proton transport is strongly related to the oxygen sublattice vibrations and has activation energies similar to (and generally somewhat lower than) those for oxygen vacancy mobility.

Proton tunnelling does not seem to play a role in proton transport in oxides at elevated temperatures.

In broad terms, the structural parameters (oxygen sublattice 'softness') that give high proton mobility also give high oxygen vacancy mobility. In the search for 'purer' proton or oxygen ion conductors it thus seems difficult to get anywhere by trying to affect defect mobilities within the same structural group. For instance, there may be little to gain beyond the present state of the art within the perovskite system in these terms. Instead, one may try to control defect concentrations, i.e. design the material to have a high or low K_3. Unfortunately for those who want proton conduction, the high K_3 that gives high proton concentrations means more densely packed oxide with 'harder' oxygen sublattices and generally lower mobilities. Systematic research on other structural systems is needed to clarify the possibility of high proton mobility combined with high proton concentrations.

REFERENCES

1. S. Stotz and C. Wagner: *Ber. Bunsenges. Phys. Chem.*, 1066, **70**(8), 781.
2. T. Norby: *Advances in Ceramics*, 1987, **23**, 107.
3. T. S. Elleman, L. R. Zumwalt and K. Verghese: *Proc. Top. Meeting Technol. Controlled Nucl. Fusion*, 1978, **3**, 1151.
4. H. Iwahara: *Solid State Ionics*, 1988, **28–30**, 573.
5. P. Kofstad: *Nonstoichiometry, Diffusion, and Electrical Conductivity in Binary Metal Oxides*, 1972. Wiley, New York.
6. T. Norby and P. Kofstad: *J. Am. Ceram. Soc.*, 1986, **69**(11), 780.
7. T. Norby and Y. Larring: *Proc. Electroceramics IV*, R. Waser (ed.), Verlag der Augustinus Buchhandlung, Aachen, ISBN 3-86073-287-0, 1994, 703.
8. T. Norby and P. Kofstad: *J. Am. Ceram. Soc.*, 1986, **69**(11), 780.
9. T. Norby, O. Dyrlie and P. Kofstad: *J. Am. Ceram. Soc.*, 1992, **75**(5), 1176.
10. Y. Larring and T. Norby: *Solid State Ionics*, 1994, **70–71**, 305.
11. Y. Larring and T. Norby: *Solid State Ionics*, 1995, **77**, 147.
12. Y. Larring, T. Norby and P. Kofstad: *Solid State Ionics*, 1991, **49**, 73.
13. T. Norby, O. Dyrlie and P. Kofstad: *Solid State Ionics*, 1992, **53–56**, 446.
14. T. Yajima and H. Iwahara: *Solid State Ionics*, 1992, **50**, 281.
15. J. F. Liu and A. S. Nowick: *Solid State Ionics*, 1992, **50**, 131.
16. T. Norby and P. Kofstad: *High Temp.-High Press.*, 1988, **20**(3), 345.
17. R. Waser: *Z. Naturforsch.*, 1987, **42a**, 1357.
18. *Handbook of Chemistry and Physics*, 69th edn, 1988–1989, CRC Press, Florida.
19. H. Yokokawa, N. Sakai, T. Kawada and M. Dokiya: *Denki Kagaku*, 1990, **58**(6), 561.
20. J. B. Bates, J. C. Wang and R. A. Perkins: *Phys. Rev. B*, 1979, **19**(8), 4130.
21. H. Iwahara, T. Hibino and T. Okada: *Ionic and Mixed Conducting Ceramics*, T. A. Ramanarayanan, W. L. Worrell and H. L. Tuller (eds), The Electrochemical Society, NJ, 1994, 1.
22. T. Norby: *Solid State Ionics*, 1988, **28–30**, 1586.
23. D. Sutija, T. Norby and P. Björnbom: *Ionic and Mixed Conducting Ceramics*, T. A. Ramanarayanan, W. L. Worrell and H. L.Tuller (eds), The Electrochemical Society, NJ, 1994, 27.
24. D. Sutija, T. Norby and P. Björnbom: *Solid State Ionics*, 1995, **77**, 167.
25. K.-D. Kreuer, E. Schönherr and J. Maier: *Solid State Ionics*, 1994, **40–41**, 278.
26. K. C. Liang and A. S. Nowick: *Solid State Ionics*, 1993, **61**, 77.
27. N. Bonanos: *Solid State Ionics*, 1992, **53–56**, 967.
28. L. Zimmermann, H. G. Bohn, W. Schilling and E. Syskakis: *Solid State Ionics*, 1995, **77**, 163.
29. R. Hempelmann, Ch. Karmonik, Th. Matzke, M. Cappadonia, U. Stimming, T. Springer and M. A. Adams: *Solid State Ionics*, 1995, **77**, 152.
30. T. Norby: *J. Phys. IV*, 1993, **3**, C9–99.
31. Yu. M. Baikov and E. K. Shalkova: *J. Solid State Chem.*, 1992, **97**, 224.
32. A. Mitsui, M. Miyayama and H. Yanagida: *Solid State Ionics*, 1987, **22**, 213.
33. N. G. Hainovsky, Yu. T. Pavlukhin and E. F. Hairetdinov: *Solid State Ionics*, 1986, **20**, 249.
34. V. V. Sinitsyn, E. G. Ponyatovski, A. I. Baranov, A. V. Tregubchenko and L. A. Shuvalov: *Sov. Phys. JETP*, 1991, **73**, 386.
35. T. Norby, M. Friesel and B.-E. Mellander: *Solid State Ionics*, 1995, **77**, 105.
36. K.-D. Kreuer, A. Fuchs and J. Maier: *Solid State Ionics*, 1995, **77**, 157.

Oxygen Diffusion in Acceptor-doped $SrCeO_3$ and $CaZrO_3$

R. A. DE SOUZA, G. GIBSON and J. A. KILNER
Department of Materials, Prince Consort Road, Imperial College of Science, Technology and Medicine, London, SW7 2BP, UK

INTRODUCTION

The aim of these experiments is to determine the oxygen diffusion coefficient in $SrCe_{0.95}Yb_{0.05}O_{3-\delta}$ (SCYb) and $CaZr_{0.9}In_{0.1}O_{3-\delta}$ (CZIn) between 500°C and 1000°C. Both the materials under investigation have been classed as proton conductors. The introduction of protons occurs via the reaction between water and oxygen vacancies.

$$V_O^{\cdot\cdot} + O_O^{\times} + H_2O \leftrightarrow 2OH_O^{\cdot} \tag{1}$$

These vacancies are produced by doping with aliovalent cations. For these compounds a trivalent cation (Yb^{3+}, In^{3+}) is substituted for the tetravalent host cation (Ce^{4+}, Zr^{4+} respectively). Oxygen may also react with the oxygen vacancies to generate electron holes according to

$$V_O^{\cdot\cdot} + \tfrac{1}{2}O_2 \leftrightarrow 2O_O^{\times} + h^{\cdot} \tag{2}$$

Characterisation of the transport properties has focused on measuring the electrical conductivity in 'wet' and 'dry' atmospheres in order to separate the individual contributions to the total conductivity.[1,2] However as oxygen vacancies are assumed to be minority carriers in the temperature range of interest, accurate and unambiguous determination of oxygen ion conductivity is not easily accomplished using this method. Use of the $^{18}O/^{16}O$ Isotope Exchange Diffusion Profile (IEDP) technique allows the oxygen diffusivity to be determined independently of other mobile species.[3] Secondary Ion Mass Spectrometry (SIMS) was employed to determine the ^{18}O diffusion profiles.

EXPERIMENTAL METHOD

Ceramic samples were polished with successive grades of diamond polish (final polish $\frac{1}{4}$ μm), prior to an equilibration anneal in $^{16}O_2$. The isotope exchange was carried out in an atmosphere of 93% enriched $^{18}O_2$ gas. The water contents of the annealing gases were measured *in situ* using a moisture meter (Shaw Ltd) and both found to be less than 2 ppm. The diffusion profiles were determined on an Atomika

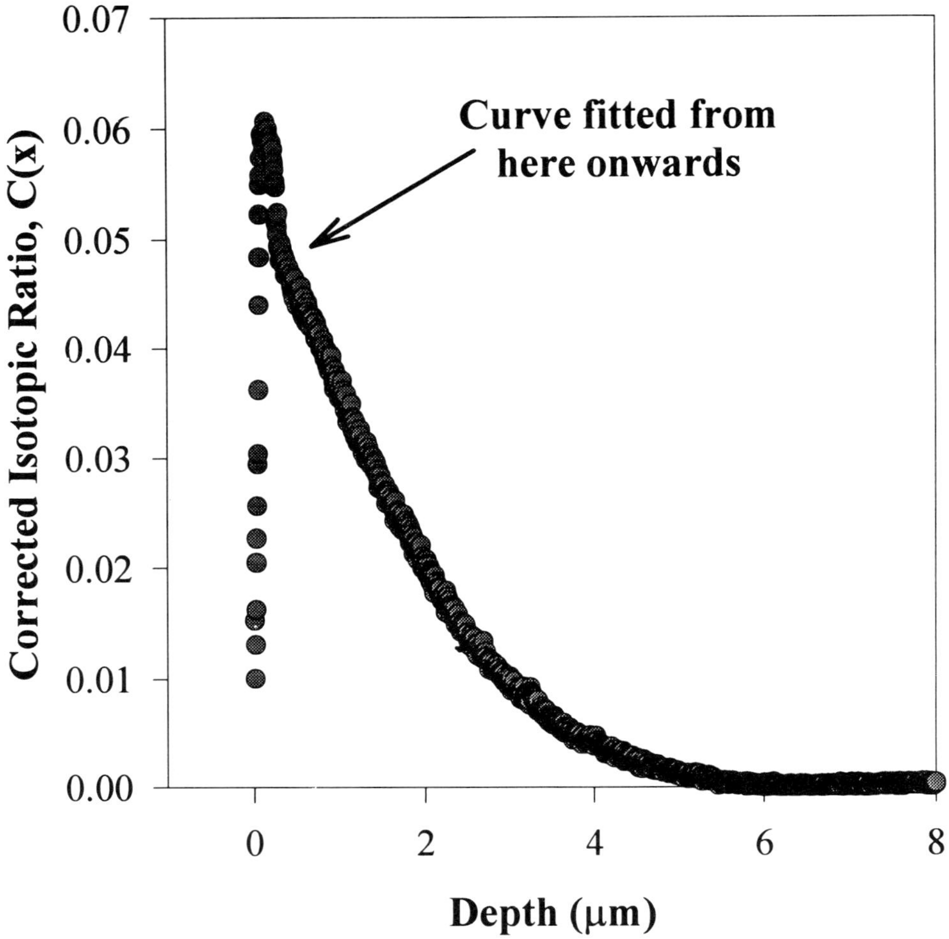

Figure 1 Diffusion profile for SCYb (^{18}O Anneal at 600°C for 520 s).

6500 SIMS instrument with a 10 keV Xe^+ or Cs^+ primary beam. A 2 kV electron beam was used to provide charge compensation. For profiles less than 10 μm the machine was operated in depth profile mode. The crater depth was measured post analysis using a surface profilometer. For penetration profiles greater than 50 μm the linescan technique was employed. Details of both are given elsewhere.[3]

The oxygen self-diffusion and surface exchange coefficients are obtained by fitting the experimental data to:

$$C(x) = \operatorname{erfc}\left[\frac{x}{2\sqrt{D_T t}}\right] - \left[\exp\left(\frac{kx}{D_T} + \frac{k^2 t}{D_T}\right) \times \operatorname{erfc}\left(\frac{x}{2\sqrt{D_T t}} + k\sqrt{\frac{t}{D_T}}\right)\right] \quad (3)$$

where D is the tracer diffusion coefficient, k is the surface exchange coefficient and t is the time of the ^{18}O isotopic anneal.

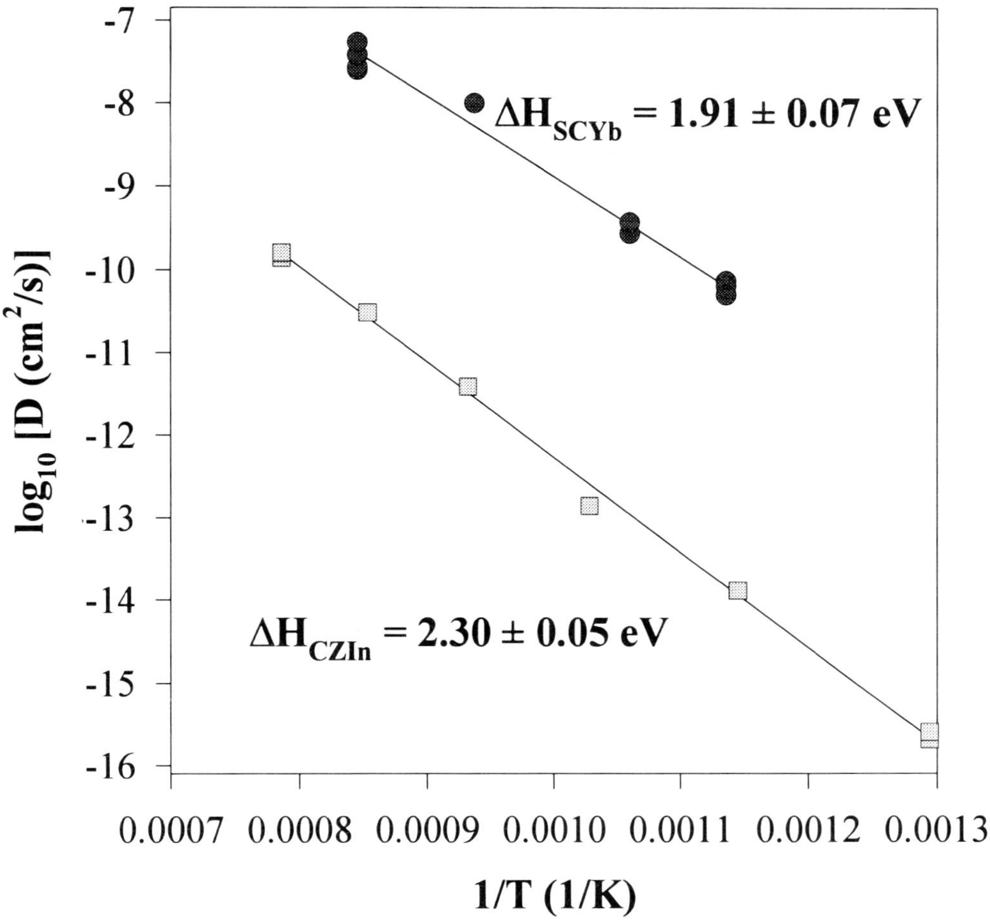

Figure 2 Oxygen Tracer Diffusion Coefficient against inverse temperature for SCYb and CZIn.

RESULTS AND DISCUSSION

A typical diffusion profile for SCYb is given in Fig. 1 for an ^{18}O anneal at 600°C. It appears that there is a dilution of the corrected isotopic ratio at the surface. In order to fit the experimental data to equation (3) it was necessary to remove the initial section and carry out the fit on the remainder of the data. Although this has negligible effect on the value of D obtained, it does result in a high degree of scatter for k and is attributed to reactivity of these ceramics with respect to $C^{16}O_2$.[4] The temperature dependence of D and k for both materials is shown in Figs 2 and 3 respectively. Arrhenius-type behaviour is assumed for both transport processes and yields the activation energies stated on the graphs. The errors given represent the

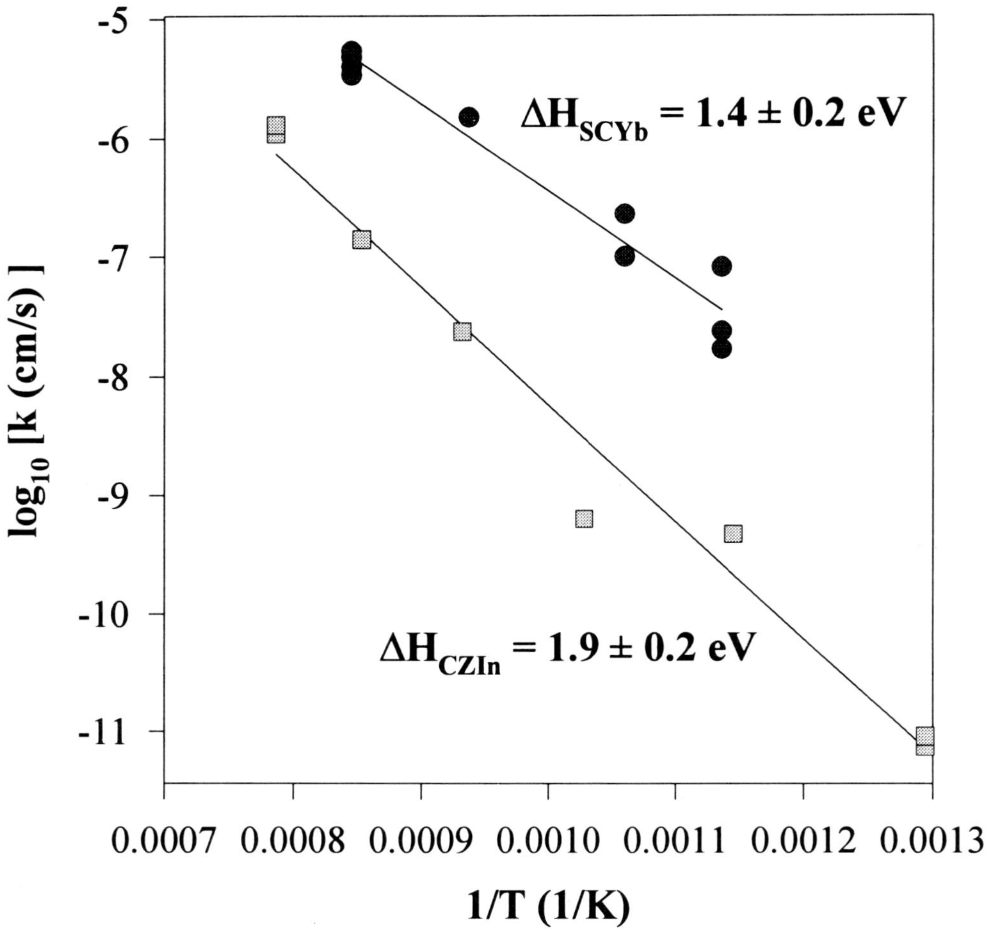

Figure 3 Oxygen Surface Exchange Coefficient against inverse temperature for SCYb and CZIn.

standard uncertainties in fitting the values of D and k to Arrhenius-type behaviour.

The activation energies for the self diffusion of oxygen in both materials are rather high: 2.30 ± 0.05 eV for CZIn and 1.91 ± 0.07 eV for SCYb. Reasonable agreement with electrical measurements is found for CZIn. Wang *et al.* reported activation energies for variously doped $CaZrO_3$ compositions from 1.8 to 3.0 eV,[5] while in a recent study by Kurita *et al.*[6] a value of 2.5 ± 0.5 eV was determined for oxygen ion conduction in $CaZr_{0.9}In_{0.1}O_3$. Unfortunately SCYb has not been subjected to the rigorous experimental approach adopted by Kurita *et al.* and hence the activation energies reported in the literature apply to proton or hole transport.[1,7]

If we consider the magnitude of the activation energies it is unlikely that these represent migration enthalpies alone as the values are so high. One possible explanation involves an additional contribution due to the trapping of oxygen vacancies, either by association between dopant cations and oxygen vacancies or by an ordering process. Here we have made the assumption that the vacancy concentration does not change in the temperature range of interest – this point will be returned to later. Evidence for an ordering of the vacancies in acceptor-doped perovskites was first suggested by the work of Knight and Bonanos.[8] Neutron powder diffraction indicated that the vacancies in Y-doped $BaCeO_3$ were confined exclusively to one of two oxygen positions, which occur in the orthorhombic structure. Ranløv *et al.*[9] have found identical results in their neutron diffraction study of $SrCeO_3$ – the oxygen vacancies are situated at apical positions of the CeO_6 octahedra. They propose that this is due to the longer Ce–O bond length (and therefore weaker electrostatic potential) associated with the apical position. It is therefore believed that the measured activation energy is the sum of ΔH_m, the oxygen vacancy migration enthalpy and ΔH_o, a term which reflects the fact that the oxygen vacancies are ordered. It is unclear at present if the degree of order is a function of temperature or whether there is a defined temperature at which an order–disorder transition occurs. In their studies on highly defective perovskites Adler *et al.*[10] also found that the vacancies were ordered, but it appears that even in lightly doped perovskites the vacancies have a tendency to order.

Now we return to our earlier assumption of an essentially constant vacancy concentration. The tracer diffusion coefficient can be expressed in terms of a vacancy diffusivity and a site fraction of vacancies. For materials such as SCYb and CZIn we would expect the vacancy concentration to be constant and equal to half the dopant concentration. Taking Yb-doped $SrCeO_3$ as an example:

$$[V_O^{\cdot\cdot}] = \tfrac{1}{2}[Yb'_{Ce}] \tag{4}$$

Lattice oxidation, that is the reaction of these vacancies with oxygen to give electron holes, would not result in vast changes in oxygen vacancy concentration. At the cross-over between the mixed-conducting and ionic domains the contributions to the conductivities are equal. Given that the mobilities of oxygen vacancies and electron holes differ by several orders of magnitude,[11] it would be expected that the concentrations would show a similar difference though in the opposite sense. The major concern, however, is the annihilation of vacancies by hydroxide ions (equation 1). Based on numerical modelling of the defect structure for SCYb,[12] our belief is that due to the low water partial pressure in which the isotope anneals were carried out, protons are minority species in the temperature range of interest (Fig. 4) and therefore the oxygen vacancy concentration is unchanged. Thus our assumption of constant vacancy concentration is considered to be correct.

The temperature above which oxygen ions dominate the ionic conductivity can be estimated by comparing electrical data with the theoretical oxygen ion conductivity calculated using the Nernst–Einstein equation:

$$\sigma = \frac{Nq^2 D}{kT} \tag{5}$$

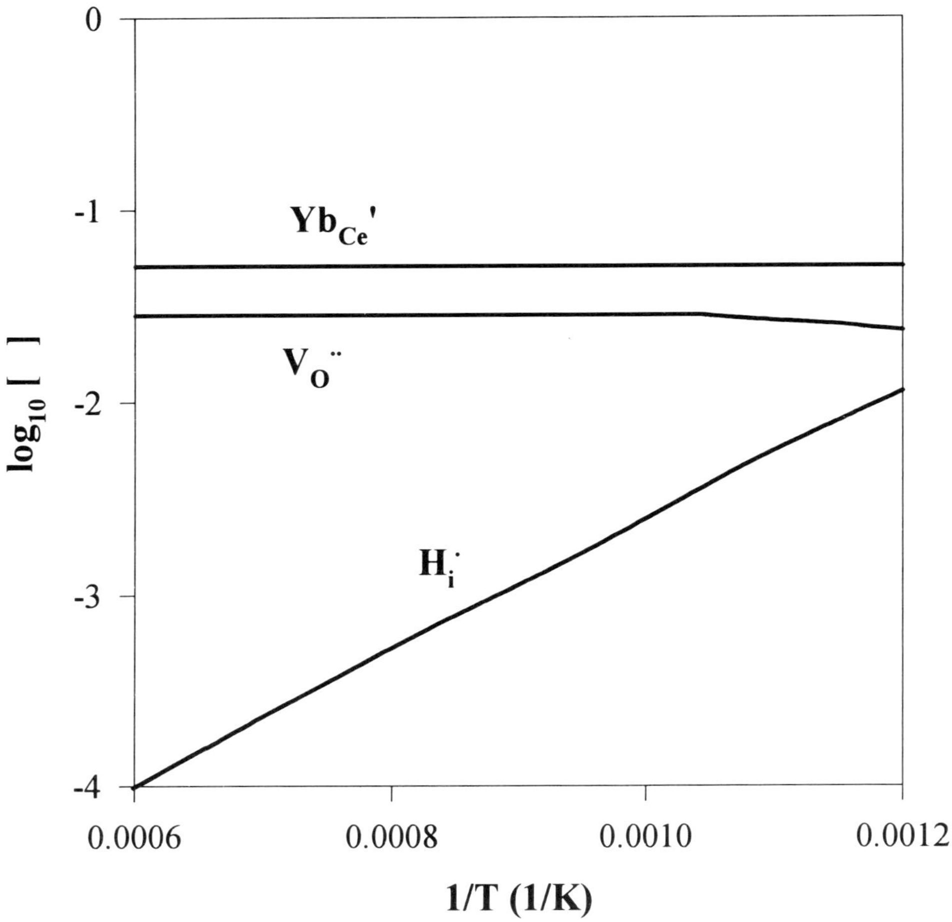

Figure 4 Calculated concentrations of oxygen vacancies and protons for SCYb as a function of temperature at $P_{H_2O} = 10^{-4.5}$.

where q, k and T have their normal meanings. N is the number of anion sites per unit volume and was determined from the lattice parameters.[9,13] It is stressed that this equation exclusively defines the relationship between σ and D; it does not require the concentration of vacancies to be included as this is contained within the transport parameters themselves. This allows the direct comparison of activation energies obtained from IEDP with those from electrical measurements.

Theoretical oxygen ion conductivities were obtained for CZIn from 500° to 1800° C using values of the oxygen diffusivities calculated from the pre-exponentials and activation energies. These are plotted in Fig. 5 together with the electrical conductivity for CZIn[13] and SCYb.[14] Data were taken from these particular references as the electrical measurements were carried out at low oxygen partial pressures and in temperature regimes in which we are confident that neither oxygen

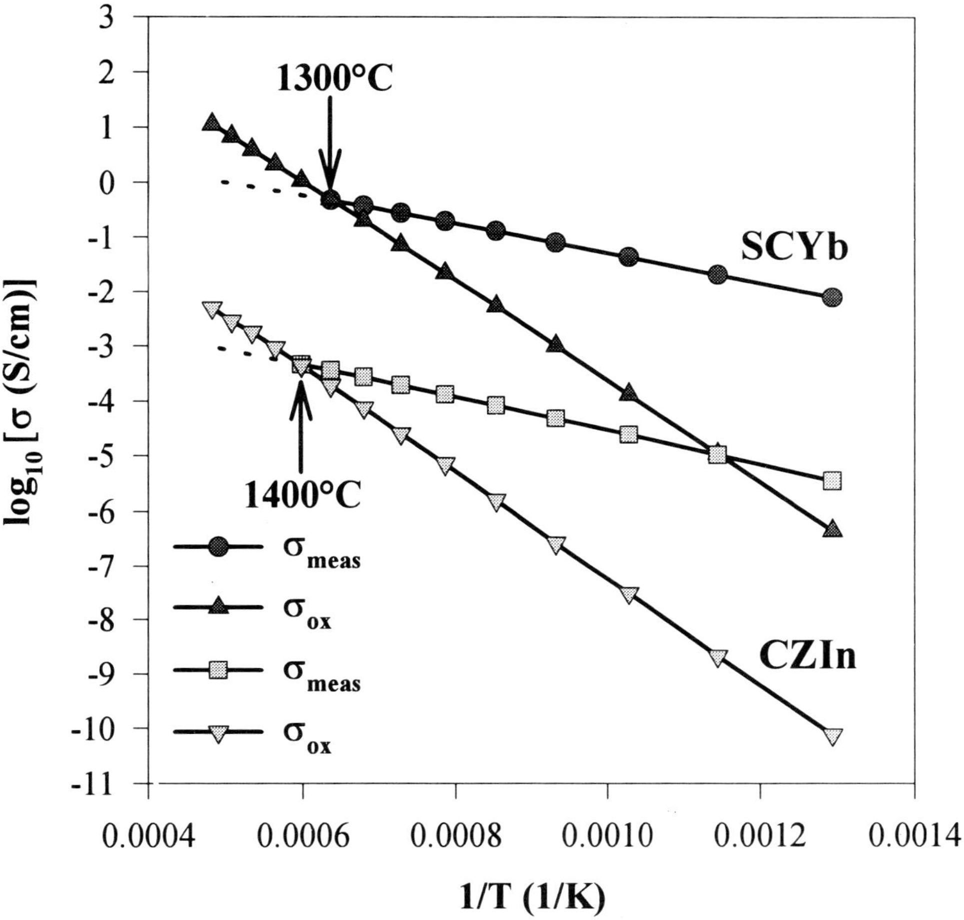

Figure 5 Extrapolated σ_{meas} and σ_{ox} showing transition temperatures to oxide ion-dominated conductivity for SCYb and CZIn.

ions nor electron holes contribute to the measured conductivity. The temperatures above which oxide ion conductivity dominates are thus estimated to be 1300°C for SCYb and 1400°C for CZIn. It is satisfying to note that Kurita *et al.*[6] reported an identical temperature for the transition to oxide ion-dominated conduction.

CONCLUSIONS

The oxygen diffusion coefficients have been determined at $P_{O_2} = 1$ atm for two acceptor-doped perovskite oxides: $CaZr_{0.9}In_{0.1}O_{3-\delta}$ and $SrCe_{0.95}Yb_{0.05}O_{3-\delta}$. Extrapolation of these results would imply that oxygen ion conduction in these materials becomes dominant above 1400°C and 1300°C respectively. It is proposed that the high activation energies obtained for oxygen transport in these materials are due to an ordered oxygen vacancy structure which reduces the concentration of 'free' or mobile vacancies.

ACKNOWLEDGEMENTS

The authors would like to thank T. Yajima, TYK Corp., Japan for supplying the samples used in this work. Financial assistance for RADS was obtained from EPSRC and Siemens AG.

REFERENCES

1. S. Shin, H. H. Huang, M. Ishigame and H. Iwahara: *Solid State Ionics*, 1991, **40/41**, 910.
2. T. Yajima, H. Kazeoka, T. Yogo and H. Iwahara: *Solid State Ionics*, 1991, **47**, 271.
3. S. Carter, A. Selcuk, R. J. Chater, J. Kajda, J. A. Kilner and B. C. H. Steele: *Solid State Ionics*, 1992, **53–56**, 597.
4. M. J. Scholten, J. Schoonman, J. C. van Miltenburg and H. A. J. Oonk: *Solid State Ionics*, 1993, **61**, 83.
5. C. Wang, X. Xu, H. Yu, Y. Wen and K. Zhao: *Solid State Ionics*, 1988, **28–30**, 542.
6. N. Kurita, N. Fukatsu, K. Ito and T. Ohashi: *J. Electrochem. Soc.*, 1995, **142**(5), 1552.
7. T. Scherban, A. S. Nowick: *Solid State Ionics*, 1989, **35**, 189.
8. K. S. Knight and N. Bonanos: *J. Mater. Chem.*, 1992, **2**(7), 709.
9. J. Ranløv, B. Lebech and K. Nielsen: *J. Mater. Chem.*, 1995, **5**(5), 743.
10. S. Adler, S. Russek, J. Reimer, M. Fendorf, A. Stacy, Q. Huang, A. Santoro, J. Lynn, J. Baltisberger and U. Werner: *Solid State Ionics*, 1994, **68**, 193.
11. E. O. Ahlgren, J. Ranløv and F. W. Poulsen: *Proc. 2nd Int. Symp. Ionic Mixed Cond. Cer.*, T. A. Ramanarayanan, W. L. Worrell and H. L. Tuller (eds), San Francisco, USA, May 1994.
12. Y. Larring: Univ. of Oslo, unpublished results.
13. G. Gibson: unpublished results.
14. N. Bonanos, B. Ellis and M. N. Mahmood: *Solid State Ionics*, 1988, **28–30**, 579.

Coulometric Titration of Ceria Solid Solutions

D. SCHNEIDER, M. GÖDICKEMEIER
and L. J. GAUCKLER
Swiss Federal Institute of Technology (ETH-Zürich), Nichtmetallische Werkstoffe, CH-8092 Zürich, Switzerland

ABSTRACT
Ceria solid solutions are potential electrolytes for solid oxide fuel cells (SOFC). At the anode side of SOFCs at low oxygen partial pressures ceria is partially reduced. The non-stoichiometry of CeO_{2-x} solid solutions with $SmO_{1.5}$, $GdO_{1.5}$ and CaO were investigated by isothermal coulometric titration in a zirconia cell in the temperature range from 700 to 900°C and in the oxygen partial pressure range from 10^{-12} to 10^{-25} atm.

The results of the three different compositions at 900°C are comparable to those for pure ceria which is more stable against reduction than ceria solid solutions at lower temperatures. The $\log x$ vs. $\log p(O_2)$ dependence suggests a defect model with oxygen vacancies and their associates with reduced cerium cations and dopant cations. Under SOFC conditions low oxygen non-stoichiometries of 0.005 to 0.008 are predicted for ceria solid solutions at 700°C.

1. INTRODUCTION

Conventional solid oxide fuel cells operating at temperatures above 900°C use zirconia electrolytes. In order to reduce operating temperatures,[1–2] ceria solid solutions were suggested as solid electrolytes[3–9] because of their higher ionic conductivity at temperatures below 800°C. However, at the anodic side, under low oxygen partial pressures, ceria is reduced and becomes non-stoichiometric. The non-stoichiometry has been measured for pure ceria,[10–15] mainly by thermogravimetry[13,15] and coulometric titration.[10]

The reduction of ceria introduces free electrons which are associated with the cerium cations. The larger ionic radii of these Ce^{3+} cations lead to an isothermal expansion which can cause detachment of the anode or cracks at the interfaces between electrolyte and electrodes.[16–20] Attempts have been unsuccessful to protect the electrolyte from reduction by coatings of thin zirconia layers,[15] due to the thermal expansion mismatch between zirconia and ceria leading to cracks in these films.

Different defect models were suggested to describe ceria solid solutions.[10,13,14] By allowing for interaction of defects, the dependence of non-stoichiometry on oxygen partial pressure can be calculated.[11]

The purpose of this study is to determine the non-stoichiometry of ceria solid solutions under SOFC operating conditions.

2. EXPERIMENTAL

2.1 Powder Preparation

Ceria solid solutions with 20 mol% of $SmO_{1.5}$ and 10 mol% of CaO were synthesised by co-precipitation. Samarium oxide (Rhône-Poulenc, 99.99%) was dissolved in hydrochloric acid and mixed with an aqueous solution of cerium nitrate (Fluka Chemie, 99.9%). The mixture was co-precipitated by dropping it slowly into ammonium oxalate solution. For the CeO_2–CaO solid solution, calcium nitrate (Merck, 99%) was used as a starting material. The precipitate was washed with water and filtered several times. After drying at 80°C, the powders were calcined at 700°C (CeO_2–Ca) or 800°C (CeO_2–Sm). The composition of the powders was verified by X-ray diffraction and electron microprobe analysis. Ceria powder with 20 mol% $GdO_{1.5}$ was supplied by Rhône-Poulenc.

2.2 Solid-state Coulometric Titration

The experimental setup with the coulometric titration cell is shown in Fig. 1. The cell consists of an alumina tube 14.5 mm in diameter and 20.8 mm high, which is closed at one end. The top of the cell is an 8 mol% yttria-stabilised zirconia electrolyte disk 1 mm thick. On both sides of the electrolyte, platinum paste (Heraeus C 3605 S) was applied. The sample powder was weighted into the cell, which was closed and sealed with glass (Degussa, Cerdec 90016). Several heating cycles up to 920°C were necessary to achieve a tight cell.

The PEN-element (Positive electrode Electrolyte Negative electrode) served as both oxygen pump and sensor. The measuring procedure consisted of two parts: first, oxygen was pumped out at a current between 5 and 15 mA and, second, the EMF (ElectroMotive Force) of this cell was followed vs time until equilibrium condition was reached. The oxygen partial pressure was calculated from the EMF

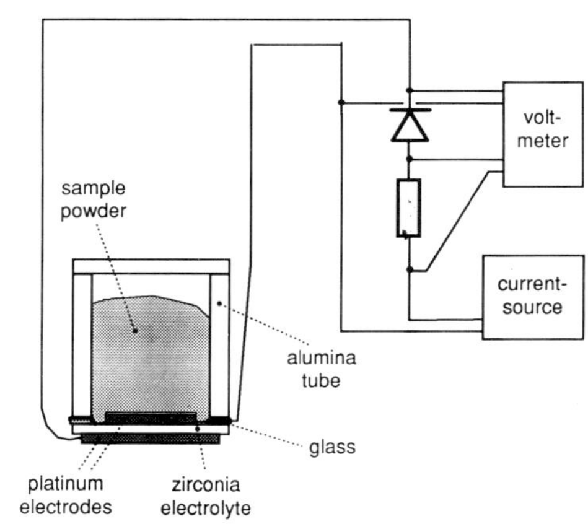

Figure 1 Coulometric titration cell and electrical equipment.

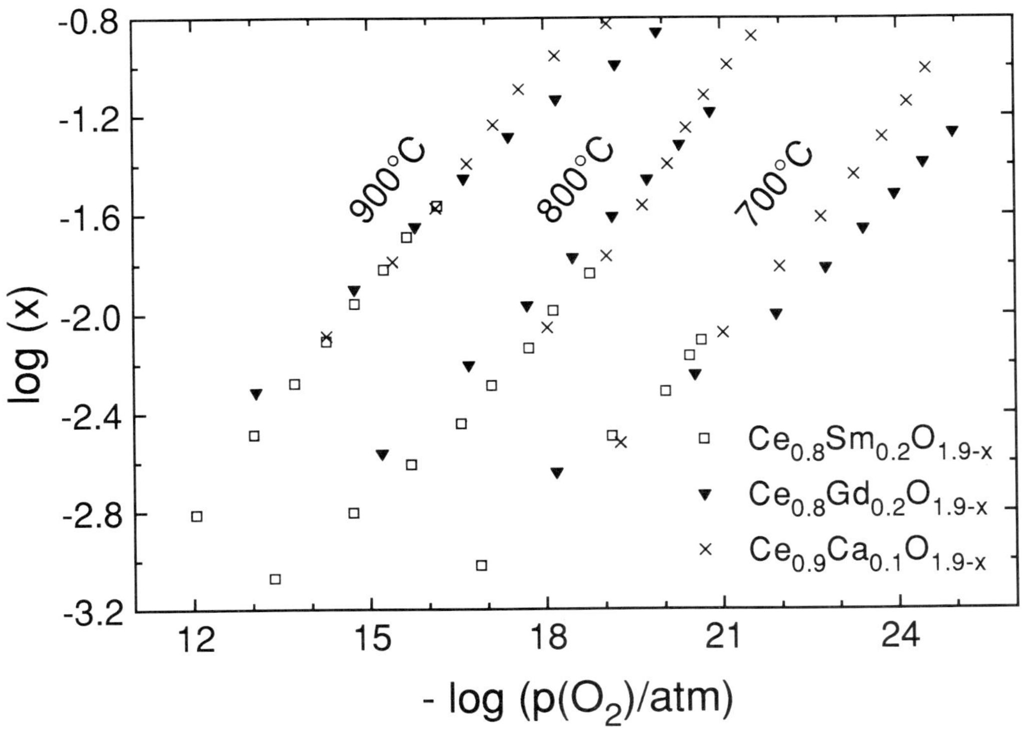

Figure 2 Dependence of non-stoichiometry of ceria solid solutions on oxygen partial pressure, measured by solid-state coulometric titration.

of the cell. The non-stoichiometry was calculated from the amount of oxygen which was pumped out of the cell. This amount was obtained by integration of the recorded cell current on the time.

Measurements were carried out with $Ce_{0.8}Sm_{0.2}O_{1.9}$, $Ce_{0.8}Gd_{0.2}O_{1.9}$ and $Ce_{0.9}Ca_{0.1}O_{1.9}$ at 700, 800 and 900°C.

3. RESULTS

The dependence of the composition upon the oxygen partial pressure was calculated from the experimental results. In Fig. 2 the results of the coulometric titration are plotted as $\log(x)$ vs $-\log(p(O_2))$. In Fig. 3 they are plotted together with literature data of pure CeO_2[13] for each temperature separately.

Among the different ceria solid solutions only small differences are visible. The difference between the ceria solid solutions and pure ceria increases with decreasing temperature. At 900°C they show the same stability against reduction but pure ceria is stronger temperature dependent. Therefore at lower temperatures, pure CeO_2 is obviously more stable against reduction, compared to ceria solid solutions.

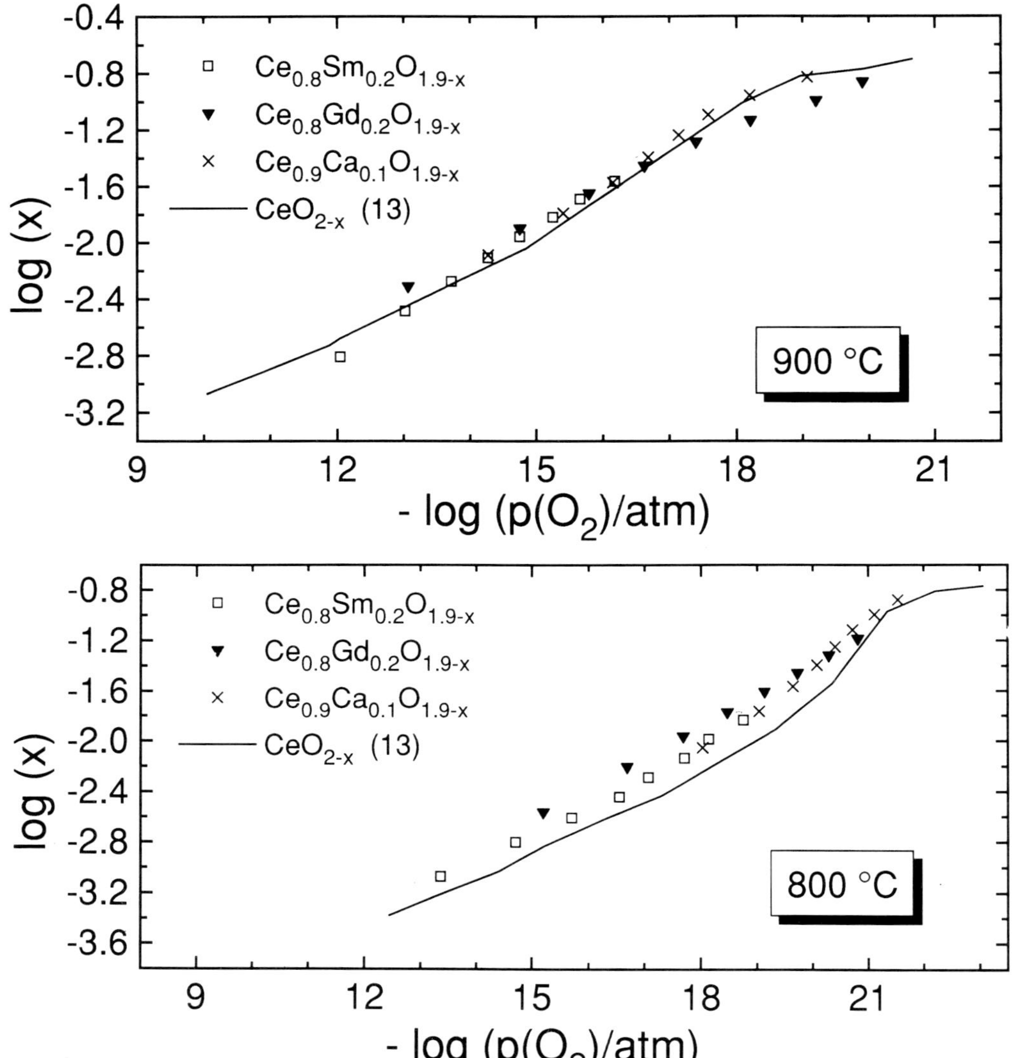

Figure 3 Dependence of non-stoichiometry of ceria solid solutions and pure ceria[13] on oxygen partial pressure at 900, 800 and 700°C.

Figure 4 shows the same results as Fig. 2. The plotted numbers indicate the reciprocal slopes corresponding to the m value in eqs (5) and (6). The slopes show m values between 3.3 and 4.5.

The curves in Figs 2–4 exhibit different shapes in the measuring range. A decreasing slope with decreasing pressure is found for ceria solid solutions with samaria and gadolinia at 800 and 900°C over the measuring range. Pure ceria and ceria-calcia solid solutions on the other hand show increasing slopes with decreasing pressure for low nonstoichiometries.

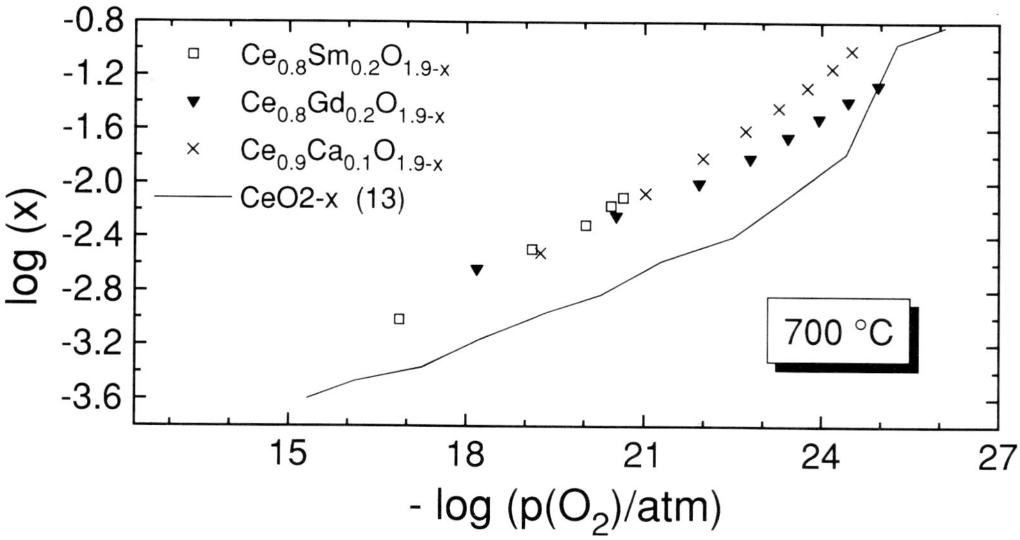

Figure 3 *continued*

4. DEFECT MODELS

In this section, the dependence of the non-stoichiometry on the oxygen partial pressure will be derived. The calculations are based, first, on a simple model describing oxygen vacancies only and, second, on a model including associated defects.

4.1 Simple Defect Model

For the following calculations, samaria and calcia can be replaced by other tri- and divalent dopant elements respectively.

The overall reduction reactions for ceria solid solutions are

$$Ce_{1-y}Ca_yO_{2-y} \leftrightarrow Ce_{1-y}Ca_yO_{2-y-x} + \frac{x}{2}O_2(g) \tag{1}$$

$$Ce_{1-y}Sm_yO_{2-y/2} \leftrightarrow Ce_{1-y}Sm_yO_{2-y/2-x} + \frac{x}{2}O_2(g) \tag{2}$$

The defect reaction in Kröger–Vink notation is (13)

$$2Ce^x_{Ce} + O^x_O \overset{K_1}{\longleftrightarrow} 2Ce'_{Ce} + V^{\cdot\cdot}_O + \frac{1}{2}O_2(g) \tag{3}$$

The mass action constant associated with the above reaction for ceria-samaria solid solutions is

$$K_1 = \frac{4x^2 \cdot (x + y/2)}{(1-y-2x)^2 \cdot (2 - y/2 - x)} \cdot p(O_2)^{1/2} \tag{4}$$

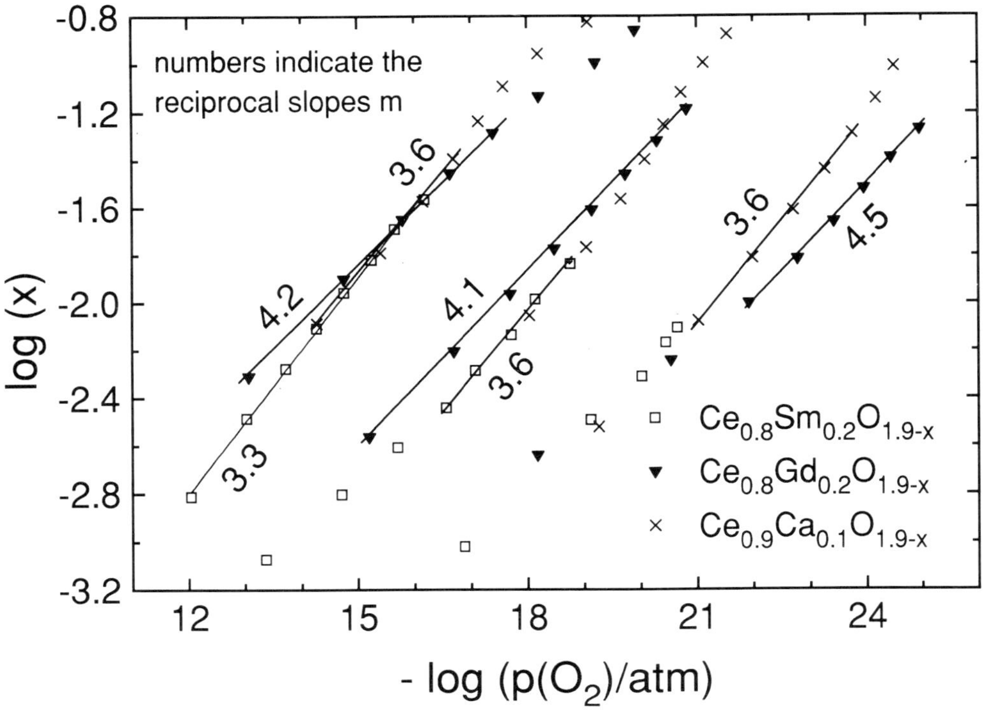

Figure 4 Reciprocal slopes of measured ($\log(x)$ vs $-\log(p(O_2))$) curves; $m = -(\mathrm{d}\log(p(O_2)))/(\mathrm{d}\log(x))$.

Simplifying the relation between $p(O_2)$ and x under the assumption that doubly ionised vacancies are dominating, we get

$$x \propto p(O_2)^{-1/m(x)} \tag{5}$$

where $m(x)$ is the reciprocal slope in the plot $\log(x)$ vs. $-\log(p(O_2))$

$$m(x) = \frac{\mathrm{d}\log(p(O_2))}{\mathrm{d}\log(x)} = 4 + \frac{2x}{x + {}^{y}/_{2}} + \frac{8x}{1 - y - 2x} + \frac{2x}{2 - {}^{y}/_{2} - x} \tag{6}$$

In Fig. 5 the above result for trivalent doped ceria is plotted together with the results for divalent ceria solid solutions and pure ceria.

The reciprocal slopes plotted in Fig. 5 indicate a decreasing slope with decreasing pressure. It cannot describe the shape of the experimental curves in the entire range studied. For small non-stoichiometries the model indicates a reciprocal slope of 4 corresponding to the experimental results.

The simple defect model is only sufficient for less reduced ceria. Now we extend the model to association of defects.

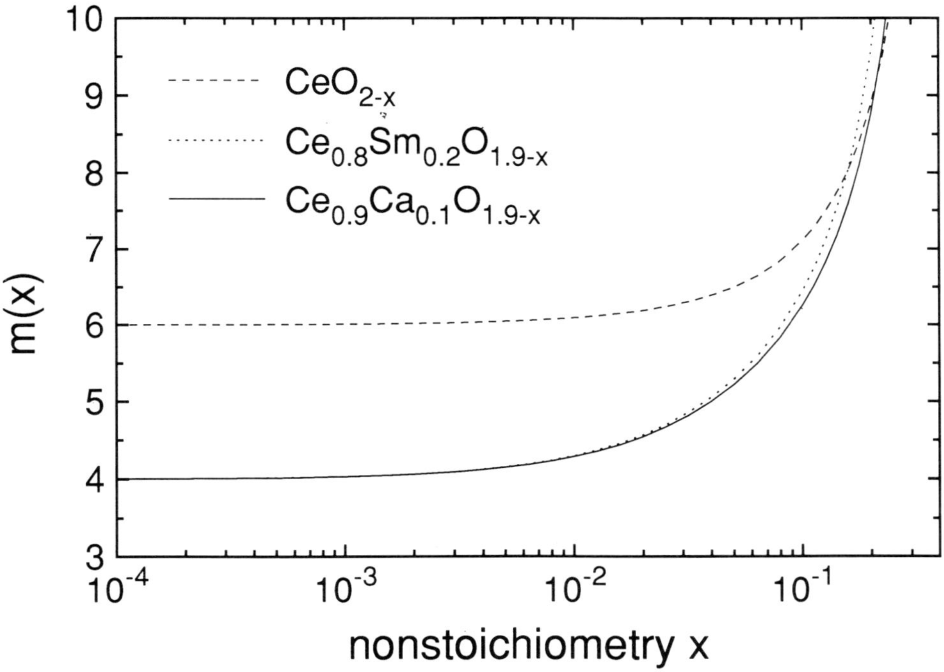

Figure 5 Calculated reciprocal slopes m dependent on non-stoichiometry for pure ceria and ceria solid solutions with di- and trivalent oxides. The calculations are based on a simple model describing oxygen vacancies only.

4.2 Associated Defects

Defects of opposite charge may form defect complexes in the ceria lattice due to their electrostatic interactions[21,22]

$$\begin{aligned}
&\mathrm{Ce}'_{\mathrm{Ce}} + V_{\mathrm{O}}^{\cdot\cdot} &&\overset{K_{4a}}{\longleftrightarrow} (\mathrm{Ce}'_{\mathrm{Ce}}\, V_{\mathrm{O}}^{\cdot\cdot})^{\cdot}\\
&\mathrm{Ce}'_{\mathrm{Ce}} + (\mathrm{Ce}'_{\mathrm{Ce}}\, V_{\mathrm{O}}^{\cdot\cdot})^{\cdot} &&\overset{K_{4b}}{\longleftrightarrow} (\mathrm{Ce}'_{\mathrm{Ce}}\, V_{\mathrm{O}}^{\cdot\cdot}\, \mathrm{Ce}'_{\mathrm{Ce}})\\
&\mathrm{Sm}'_{\mathrm{Ce}} + V_{\mathrm{O}}^{\cdot\cdot} &&\overset{K3_{a}}{\longleftrightarrow} (\mathrm{Sm}'_{\mathrm{Ce}}\, V_{\mathrm{O}}^{\cdot\cdot})^{\cdot}\\
&\mathrm{Sm}'_{\mathrm{Ce}} + (\mathrm{Sm}'_{\mathrm{Ce}}\, V_{\mathrm{O}}^{\cdot\cdot})^{\cdot} &&\overset{K_{3b}}{\longleftrightarrow} (\mathrm{Sm}'_{\mathrm{Ce}}\, V_{\mathrm{O}}^{\cdot\cdot}\, \mathrm{Sm}'_{\mathrm{Ce}})\\
&\mathrm{Ca}''_{\mathrm{Ce}} + V_{\mathrm{O}}^{\cdot\cdot} &&\overset{K_{2}}{\longleftrightarrow} (\mathrm{Ca}''_{\mathrm{Ce}}\, V_{\mathrm{O}}^{\cdot\cdot})
\end{aligned} \tag{7–11}$$

where K_i are the appropriate equilibrium constants.

The following calculations apply to calcia doped ceria. The electroneutrality condition, the dopant concentration and of the reduced cerium cations are expressed by

$$2 \cdot [\mathrm{Ca}''_{\mathrm{Ce}}] + [\mathrm{Ce}'_{\mathrm{Ce}}] = [(\mathrm{Ce}'_{\mathrm{Ce}} V_{\mathrm{O}}^{\cdot\cdot})^{\cdot}] + 2 \cdot [V_{\mathrm{O}}^{\cdot\cdot}] \tag{12}$$

$$[\mathrm{Ca}''_{\mathrm{Ce}}] + [(\mathrm{Ca}''_{\mathrm{Ce}} V_{\mathrm{O}}^{\cdot\cdot})] = y \tag{13}$$

$$[\mathrm{Ce}'_{\mathrm{Ce}}] + [(\mathrm{Ce}'_{\mathrm{Ce}} V_{\mathrm{O}}^{\cdot\cdot})^{\cdot}] + 2 \cdot [(\mathrm{Ce}'_{\mathrm{Ce}} V_{\mathrm{O}}^{\cdot\cdot} \mathrm{Ce}'_{\mathrm{Ce}})] = 2x \tag{14}$$

From eqs 7–14 a system of 3 equations is obtained

$$2 \cdot [\mathrm{Ca}''_{\mathrm{Ce}}] + [\mathrm{Ce}'_{\mathrm{Ce}}] = K_{4a} \cdot [\mathrm{Ce}'_{\mathrm{Ce}}] \cdot [V_{\mathrm{O}}^{\cdot\cdot}] + 2 \cdot [V_{\mathrm{O}}^{\cdot\cdot}] \tag{15}$$

$$[\mathrm{Ca}''_{\mathrm{Ce}}] + K_2 \cdot [\mathrm{Ca}''_{\mathrm{Ce}}] \cdot [V_{\mathrm{O}}^{\cdot\cdot}] = y \tag{16}$$

$$[\mathrm{Ce}'_{\mathrm{Ce}}] + K_{4a} \cdot [\mathrm{Ce}'_{\mathrm{Ce}}] \cdot [V_{\mathrm{O}}^{\cdot\cdot}] + 2K_{4a} K_{4b} \cdot [\mathrm{Ce}'_{\mathrm{Ce}}]^2 \cdot [V_{\mathrm{O}}^{\cdot\cdot}] = 2x \tag{17}$$

For given parameters K_i, y and x this system can be solved and with eq. 3 the dependence of $\log(p(\mathrm{O}_2))$ on $\log(x)$ is expressed as

$$\log(p(\mathrm{O}_2)) - 2 \log K_1 = -2 \log \left(\frac{[\mathrm{Ce}'_{\mathrm{Ce}}] \cdot [V_{\mathrm{O}}^{\cdot\cdot}]}{(1 - y - 2x)^2 \cdot (2 - y - x)} \right) \tag{18}$$

Equivalent equations for pure ceria as well as ceria solid solutions with trivalent oxides can be calculated analogously. Calculations for pure ceria and ceria solid solutions with di- and trivalent oxides with different K_i are shown in Fig. 6.

The shape of calculated curves depends on K_i corresponding to the association energies. The curves show slopes of 1/4 for a nearly stoichiometric lattice, followed by decreasing slopes for low pressures. For higher K_i values the shape is similar to those of ceria-calcia solid solutions and pure ceria.

5. CONCLUSIONS

The non-stoichiometry of ceria solid solutions under SOFC operating conditions (10^{-20} atm, 700°C) is about 0.005 corresponding to a linear isothermal expansion of the lattice of 0.05%[23] which seems tolerable for SOFC application. It even may be speculated that a slight reduction causes a useful electronic conductivity on the surface at the anode side of the electrolyte.

The dependence of non-stoichiometry on oxygen partial pressure deviates from −1/4, a value which would be expected for the assumption of no interactions between defects. At temperatures between 700 and 900°C associate complexes have to be assumed. The corresponding association energies were found to be higher for doping with divalent cations than with trivalent cations. The temperature depend-

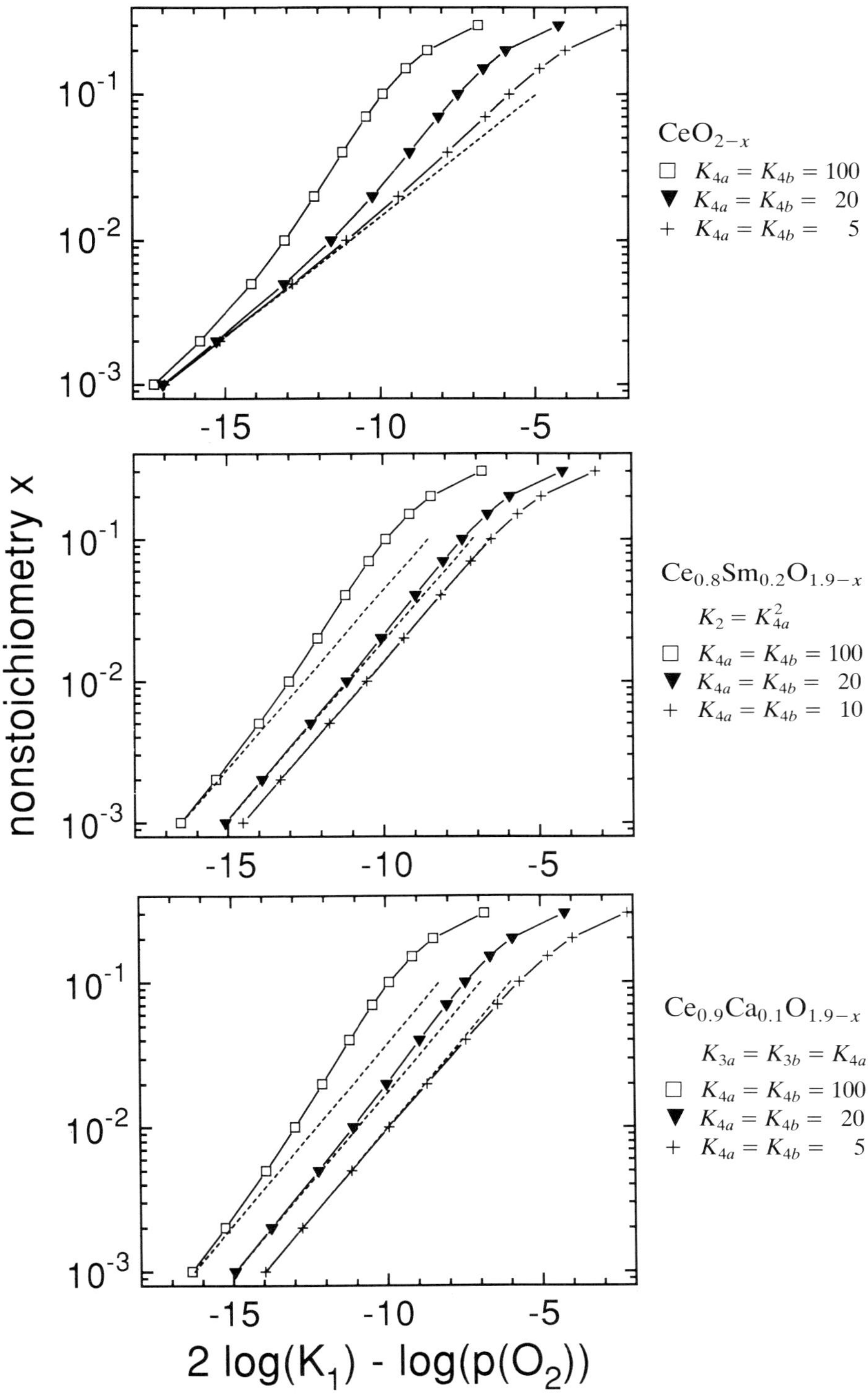

Figure 6 Calculated plots based on a defect model with associated defects. The dotted lines show the $-1/6$ and the $-1/4$ dependence of $\log(x)$ on $\log(p(O_2))$ for pure ceria and ceria solid solutions respectively.

ence of the reducibility was found to be higher for pure ceria than for ceria solid solutions. Further studies are needed to explain this surprising result.

ACKNOWLEDGEMENTS

This work was supported by the Priority Programme on Materials of the board of the Swiss Federal Institutes of Technology.

REFERENCES

1. K. Krist, M. Karpuk and J. Wright: Fuel Cell Seminar, 1992, Tucson (USA) 624.
2. H. Altdorfer, K. Honegger: *Proc. 1st Swiss Conference on Materials Research for Engineering Systems*, 1994, Technische Rundschau, Sion, Switzerland, 56.
3. D. Singman: *J. Electrochem. Soc.*, 1966, **113**, 502.
4. R. N. Blumenthal, F. S. Burgner and J. E. Garnier: *J. Electrochem. Soc.*, 1973, **120**, 1230.
5. H. L. Tuller and A. S. Nowick: *J. Electrochem. Soc.*, 1975, **122**, 255.
6. T. Kudo and Y. Obayashi: *J. Electrochem. Soc.*, 1976, **123**, 415.
7. R. T. Dirstine, R. N. Blumenthal and T. F. Kuech: *J. Electrochem. Soc.*, 1979, **126**, 264.
8. M. D. Hurley and D. K. Hohnke: *J. Phys. Chem. Solids*, 1980, **41**, 1349.
9. M. Gödickemeier, K. Sasaki, P. Bohac and L.J. Gauckler: *Advanced Solid Oxide Fuel Cells*, Proc. 6th IEA Workshop on SOFC, ENEA, 1994, Rome, Italy, 225.
10. M. A. Panhans and R. N. Blumenthal: *Solid State Ionics*, 1993, **61**, 279.
11. H. L. Tuller and A. S. Nowick: *J. Electrochem. Soc.*, 1979, **126**, 209.
12. J. W. Dawicke and R. N. Blumenthal: *J. Electrochem. Soc.*, 1988, **133**, 904.
13. R. J. Panlener, R. N. Blumenthal and J. E. Garnier: *J. Phys. Chem. Solids*, 1975, **36**, 1213.
14. D. J. M. Bevan and J. Kordis: *J. Electrochem. Soc.*, 1964, **26**, 1509.
15. K. Eguchi, T. Setoguchi, T. Inoue and H. Arai: *Solid State Ionics*, 1992, **52**, 165.
16. B. C. H. Steele: *Proc. 4th International Conference on Electronic Ceramics & Applications*, Aachen, Germany, 1994, 663.
17. M. Mogensen, T. Lindegaard, U.R. Hansen and G. Mogensen: *J. Electrochem. Soc.*, 1994, **141**, 2122.
18. G. Mogensen and M. Mogensen: *Thermochimica Acta*, 1993, **214**, 47.
19. K. Mehta: *Proc. 3rd International Symposium on SOFC*, S. C. Singhal and H. Iwahara (eds), 1994, 92.
20. B. Zachau-Christiansen, T. Jacobsen, K. West and S. Skaarup: *Proc. 3rd International Symposium on SOFC*, S. C. Singhal and H. Iwahara (eds), 1993, 104.
21. W. D. Kingery, H. K. Bowen and D. R. Uhlmann: *Introduction to Ceramics*, 2nd, 1976, John Wiley & Sons, New York, 148.
22. S. Ling: *Physical Review B*, 1994, **49**, 864.
23. Huann-Wu Chiang, R. N. Blumenthal and R. A. Fournelle: *Solid State Ionics*, 1993, **66**, 85.

Crystal Structure, Phase Relations and Oxygen Nonstoichiometry in Perovskite-type Oxide, $La_{1-x}Sr_xMnO_3$

HIROAKI TAGAWA, JUNICHIRO MIZUSAKI*, HIDETAKA NAMBU, CHITOSE NAKAO, HIROSHI TAKAI AND HIDEHIRO MINAMIUE†

Institute of Environmental Science and Technology, Yokohama National University, 156 Tokiwadai, Hodogaya-Ku, Yokohama, 240, Japan

**Research Institute for Scientific Measurements, Tohoku University 2-1-1 Katahira, Aoba-Ku, Sendai, 980-77, Japan*

†Central Research Laboratory, Kyocera Corporation, Yamashita-Cho, Kokubu, Kagoshima, 899-43, Japan

ABSTRACT

The crystal structure of a perovskite-type oxide, $La_{1-x}Sr_xMnO_3$ ($x = 0, 0.1, 0.3$) was determined as a function of temperature and oxygen pressure by high-temperature X-ray diffraction, and the phase relations in the system were derived. Further the oxygen nonstoichiometry in the oxide of $x = 0$, 0.2 and 0.4 was examined in gravimetry and coulometric titration at 600–1000°C in 1–10^{-25} atm O_2. Four modifications were found in $La_{1-x}Sr_xMnO_3$ ($x = 0$–0.3): Cubic, rhombohedral, two orthorhombic structure. The crystal structure depends on not only temperature, but also oxygen pressure and Sr content. The structure changes from orthorhombic to cubic through rhombohedral with increasing Sr content and temperature. The oxygen nonstoichiometry in $La_{1-x}Sr_xMnO_3$ with different Sr content is depicted as an oxygen content vs. $\log p_{O_2}$ curve. The curve is divided into three composition regions: hyperstoichiometric ($d > 0$), stoichiometric ($d = 0$), and hypostoichiometric ($d < 0$). As oxygen pressure is reduced, oxygen content in the oxide decreases to make plateau with the stoichiometry of $La_{1-x}Sr_xMnO_3$, where the oxidation state of Mn is +3 at $x = 0$ and increases with doped Sr content, then the oxygen content again goes down to the composition when the decomposition occurs. The decomposition temperature is affected by doped Sr content.

1. INTRODUCTION

A perovskite-type oxide, $La_{1-x}Sr_xMnO_{3\pm d}$, or denoted as $(La,Sr)MnO_3$, is used as a cathode material in solid oxide fuel cells (SOFC), which is operated around 1000°C. The surface of a cathode oxide is put in air and oxygen pressure decreases toward the electrolyte. It is thought that the oxygen pressure becomes about 10^{-3} or 10^{-4} atm on the interface between the cathode and the electrolyte. Therefore, the knowledge on the crystal structure, thermal expansion coefficient, phase relations and thermodynamic stability of $(La,Sr)MnO_3$ are required as a function of temperature and oxygen pressure. These properties at high temperatures in a wide range of oxygen pressures, however, have been scarcely known, as described in Mizusaki's review.[1]

We have been interested how crystal structure, phase relations and thermodynamic properties, nonstoichiometry, chemical reactivity and electrical properties of perovskite-type oxides with transition metals in B sites such as Cr, Mn, Fe, Co and Ni are affected by temperature and oxygen pressure. In the present work, crystal structure and nonstoichiometry in $(La,Sr)MnO_3$ including $LaMnO_3$ were examined as a function of temperature and oxygen pressure to clarify the phase relations under SOFC operating conditions.

2. EXPERIMENTAL

2.1 Samples

Precursors of $(La,Sr)MnO_3$ were prepared from mixtures of aqueous acetate solutions of component metals, La, Sr, and Mn by freeze-drying.[2] The resultant dried and mixed acetates were calcined to convert the oxides at 1000°C, then again fired at 1400°C in air after well mixed. The content of constituent metals in precursors and resultant oxides was analysed by inductively coupled plasma atomic emission spectrometer (ICP, Seiko Instruments Inc., Japan). The oxides were also confirmed to be single phase in X-ray diffraction (XRD).

2.2 High-temperature X-ray Diffraction (XRD) Measurements

A Pt flat sample holder, which was contained in a vacuum-tight chamber with an electric heater of Pt + 40%Rh wire, was attached to an X-ray diffractometer (RIGAKU Co., Japan). Oxide powders were mounted onto the sample holder. Sample temperature was changed between room temperature and 1000°C, where the temperature was measured by Pt/Pt + 13%Rh thermocouple attached on the sample holder. Oxygen pressure above 10^{-5} atm in the chamber was prepared by changing flow rates of oxygen and argon, and oxygen pressure below 10^{-8} atm by changing in H_2O-to-H_2 ratio in flowing hydrogen saturated with water vapour. Oxygen pressure was monitored by an yttria-stabilised zirconia (YSZ) oxygen sensor set at the outlet of the furnace chamber to examine whether the sample was in equilibrium.

XRD measurements were made using CuKα radiation by means of step scanning method, in width of 0.04°/2s for 3–100° or 0.04°/20s for 60–100° as 2θ. Most measurements were carried out by changing in temperature under the condition that a gas mixing ratio of O_2/Ar or H_2O/H_2 was kept constant. The measurements below 700°C was carried out at a lowered, desired temperature after a sample was kept in equilibrium at 700°C. It is expected that the oxygen stoichiometry of the oxide is kept constant below 700°C even if temperature is quickly lowered from 700°C. Observed diffraction patterns were analysed in the Rietveld method using a computer program, RIETAN, composed by Izumi,[3] and crystal structure and space groups and lattice parameters were determined.

2.3 Thermal Balance Measurements

Oxygen nonstoichiometry was *in situ* determined in thermogravimetry by using an electrobalance (Cahn Type-1000, USA) as a thermal balance in flowing gas with controlled various oxygen pressures. Oxygen pressure was changed in flow rates of oxygen and argon, or in an H_2O/H_2 mixing ratio in flowing hydrogen with water vapour, and monitored by another zirconia oxygen sensor by similar way as carried out in high-temperature XRD.

Coulometric titration was also performed for examining the nonstoichiometry in the oxide below 10^{-3} atm O_2. The data obtained were connected with the result by the thermal balance above 10^{-5} atm O_2. A sample was contained and sealed in a YSZ tube closed one end, where porous Pt electrodes were put on inside and outside walls. The oxygen content was varied by the method that current was applied from inside to outside or vice versa. The equilibrium oxygen pressure was determined by EMF measurements between both the electrodes.

3. RESULTS AND DISCUSSION

3.1. Crystal Structure and Phase Relations

3.1.1 $LaMnO_3$

The crystal structure of $LaMnO_3$ at room temperature, where the oxide is rigorously hyperstoichiometric $LaMnO_{3+d}$, is expected to be rhombohedral or orthorhombic on the basis of Galasso's consideration on ionic radii of metals, A and B, in perovskite-type oxides, $A^{3+}B^{3+}O_3$.[4] Actually, the crystal structure with La^{3+} and Mn^{3+} was rhombohedral ($R\bar{3}C$) at temperatures between room temperature and 1000°C in oxygen pressures between 1 and at least 10^{-5} atm. When oxygen pressure was further reduced, two orthorhombic structures, Pbnm(II) and Pbnm(III), were observed below 700°C. As a result, it was confirmed that there are three modifications in $LaMnO_3$, as observed for quenched samples to room temperature by Takeda *et al.*[5]

Phase diagrams are composed by drawing the data of the crystal structure, obtained *in situ* in the present work, on temperature and oxygen pressure plane. Figure 1 shows the phase diagram for $LaMnO_3$ as a function of temperature and oxygen pressure. It is seen that the crystal structure is related to the oxygen nonstoichiometry which changes the oxidation states of Mn and ionic radii of component metals, La^{3+}, Mn^{3+} and Mn^{4+}.

3.1.2 $(La,Sr)MnO_3$

When Sr is doped in $LaMnO_3$, where the oxide is denoted as $(La,Sr)MnO_3$, the relationship between crystal structure and temperature and oxygen pressure of the oxide is slightly changed from $LaMnO_3$. The crystal structure of $La_{0.9}Sr_{0.1}MnO_3$ is drawn as the phase diagram Fig. 2. Here, the structure is only rhombohedral at least under the experimental conditions. When Sr was further doped to $x = 0.3$, it was

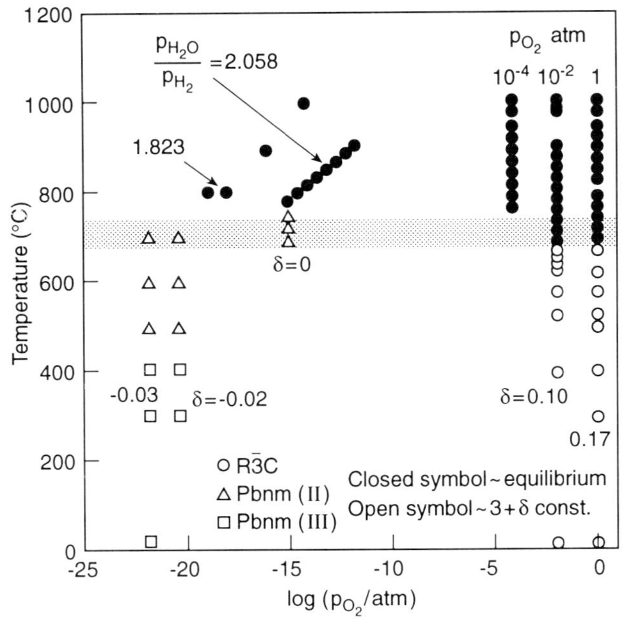

Figure 1 Variation in crystal structure of $LaMnO_3$ as a function of temperature and oxygen pressure.

found that a cubic structure is newly formed at higher temperatures in addition to the rhombohedral. The phase diagram of $La_{0.7}Sr_{0.3}MnO_3$ is shown in Fig. 3.

As shown in the phase diagrams of Figs 1–3, the crystal structure depends on not only temperature and oxygen pressure, but also doped Sr content. The structure changes as orthorhombic, rhombohedral and cubic in the order as doped Sr content increases. By partly replacing La^{3+} of 1.045 Å with Sr^{2+} of 1.13 Å where Shannon and Prewitt's ionic radius[4] is used, the oxidation state and ionic radius for Mn^{3+} of 0.58–0.645 Å are changed in Mn^{4+} of 0.54 Å. As mean ionic radius of A metals is larger than that of the original La ion by doping Sr, and that of B metals is lowered, the distortion of the structure is relaxed by doping Sr, and the crystal symmetry is raised to be cubic.

3.2 Variation in Lattice Parameters

3.2.1 $LaMnO_3$

The variation in the lattice parameters of rhombohedral $LaMnO_3$ are shown in Fig. 4 as a function of temperature and oxygen pressure. Here, the data on the lattice parameters were obtained under a constant oxygen pressure above 700°C, where the oxygen content, $3 + d$, in the oxide is varied with temperature, as later described. The lattice parameters below 700°C show those of the oxide with the oxygen content which is frozen at about 700°C. The variation in the lattice parameters with the same oxygen content above 700°C should be re-calculated by using the nonstoichiometric

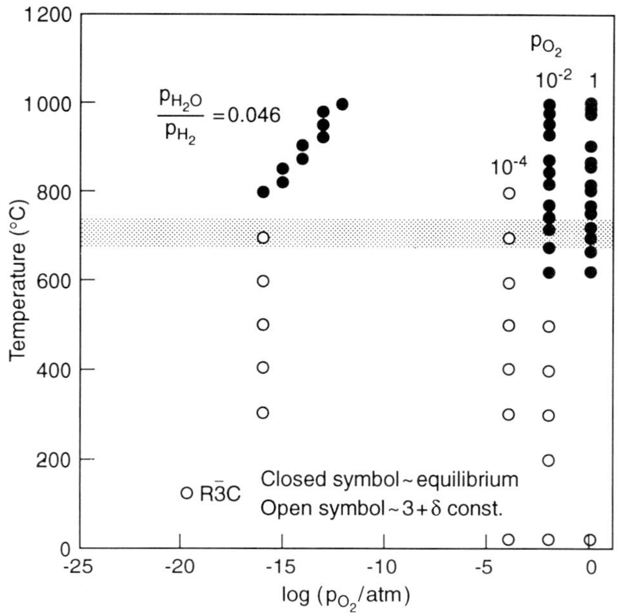

Figure 2 Variation in crystal structure of $La_{0.9}Sr_{0.1}MnO_3$ as a function of temperature and oxygen pressure.

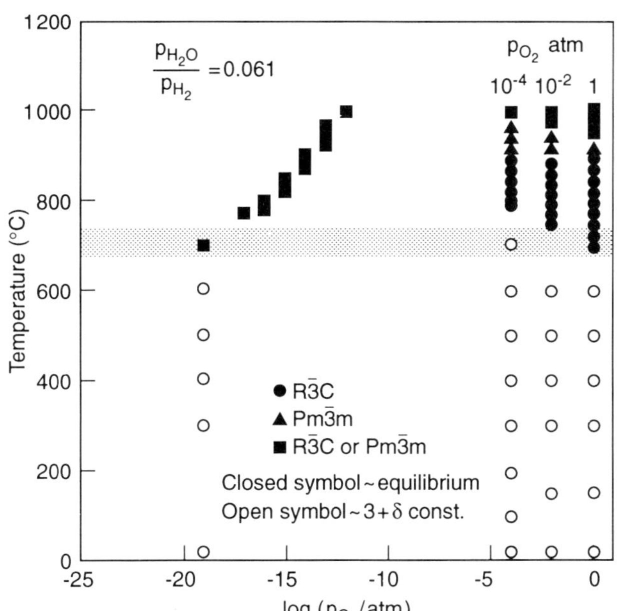

Figure 3 Variation in crystal structure of $La_{0.7}Sr_{0.3}MnO_3$ as a function of temperature and oxygen pressure.

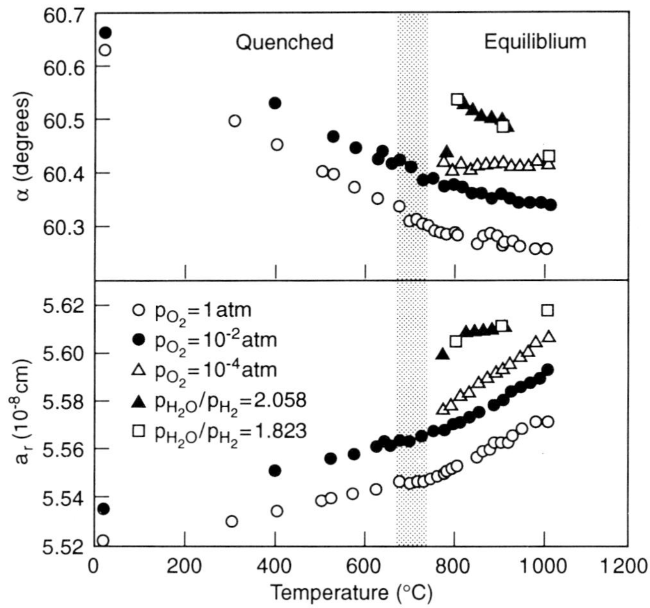

Figure 4 Lattice parameters of rhombohedral $LaMnO_3$ as a function of temperature and oxygen pressure.

data when the defect structure is rigorously considered.

In order to compare the lattice expansion of $LaMnO_3$ with that of $(La,Sr)MnO_3$ with different Sr content, the volume of the unit cell is used in the present work. The variation in unit cell volume of rhombohedral $LaMnO_3$, calculated from Fig. 4, is shown in Fig. 5. The figure shows that the cell volume increases as oxygen pressure decreases. This means that increasing in mean Mn ionic radius due to lowering in the mean Mn valence makes increase in the unit cell volume, while increasing in oxygen defects will cause the decrease in the volume, and the former overcomes the latter as a result.

3.2.2 (La,Sr)MnO$_3$

The variation way in the lattice parameters of rhombohedral $La_{0.9}Sr_{0.1}MnO_3$ as a function to temperature and oxygen pressure is similar to that of $LaMnO_3$. The unit cell volumes calculated from the lattice parameters are shown in Fig. 6. The cell volume has a tendency to increase as oxygen pressure decreases.

$La_{0.7}Sr_{0.3}MnO_3$, as given in Fig. 3, has two modifications: rhombohedral and cubic. The rhombohedral structure exists from room temperature to 900°C in $1–10^{-10}$ atm O_2 and changes into the cubic above 900°C. When oxygen pressure is reduced below 10^{-10} atm O_2, the phase transition temperature of the rhombohedral to the cubic is slightly lowered.

The unit cell volume for $La_{0.7}Sr_{0.3}MnO_3$, calculated from the lattice parameters,

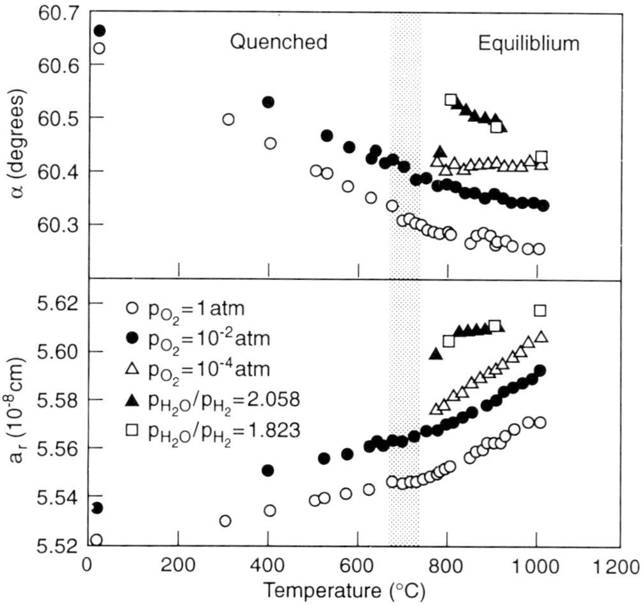

Figure 5 Unit cell volume of rhombohedral $LaMnO_3$ as a function of temperature and oxygen pressure.

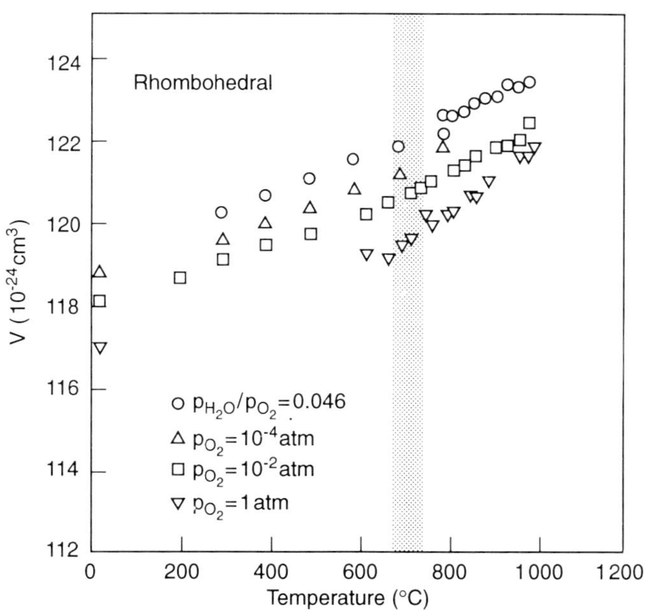

Figure 6 Unit cell volume of rhombohedral $La_{0.9}Sr_{0.1}MnO_3$ as a function of temperature and oxygen pressure.

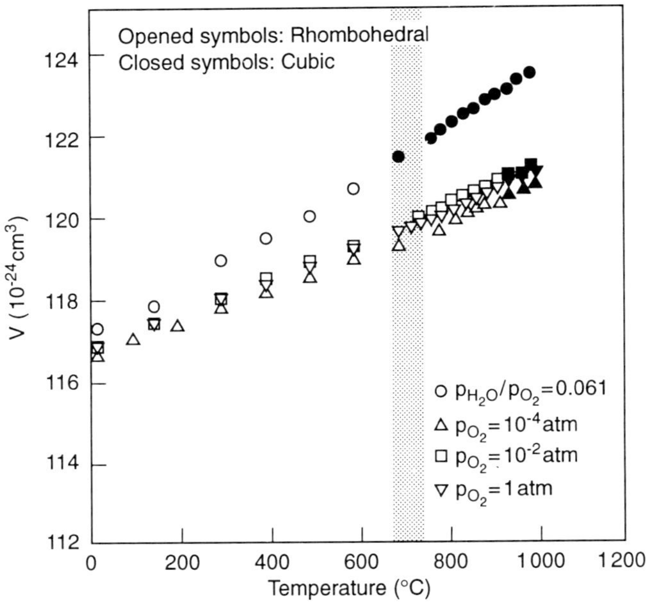

Figure 7 Unit cell volume of cubic and rhombohedral $La_{0.7}Sr_{0.3}MnO_3$ as a function of temperature and oxygen pressure.

has continuity below and above the phase transition temperature, as shown in Fig. 7. The cell volumes in oxygen pressures between 1 and 10^{-4} atm give nearly the same values regardless of oxygen pressure. This shows the result that volume expansion due to increasing in the mean Mn radius is compensated with volume reduction due to increasing in the oxygen defects.

3.3 Nonstoichiometry

It is known that the perovskite-type oxide, (La,AE)MnO_3, where AE is Ca and Sr, has a wide range of oxygen nonstoichiometry, which affects the material's stability and physical and chemical properties under operating conditions of SOFC. Although measurements of the nonstoichiometry have been performed by some workers, there is no systematic study as a function of dopant content.

The oxygen nonstoichiometry in $La_{1-x}Sr_xMnO_3$ ($x = 0$, 0.2, 0.4) is given as oxygen content vs. log p_{O_2} curves at temperatures between 600 and 1000°C in Figs 8, 9 and 10, respectively. As seen in the figures, oxygen content in all curves is divided into three regions in relation to oxygen pressure: hyperstoichiometric ($d > 0$), stoichiometric ($d = 0$) and hypostoichiometric ($d < 0$). When oxygen pressure is reduced from 1 atm, the oxygen content, $3 + d$, in the oxides decreases to a constant value, then is kept constant to make a plateau for some decade of oxygen pressure, and again goes down to the decomposition pressure. The oxygen content at 600°C in 1 atm O_2, which was obtained by the thermal balance, was 3.16, 3.06 and 3.02 for $x = 0$, 0.2 and 0.4, respectively.[6] These values were in good

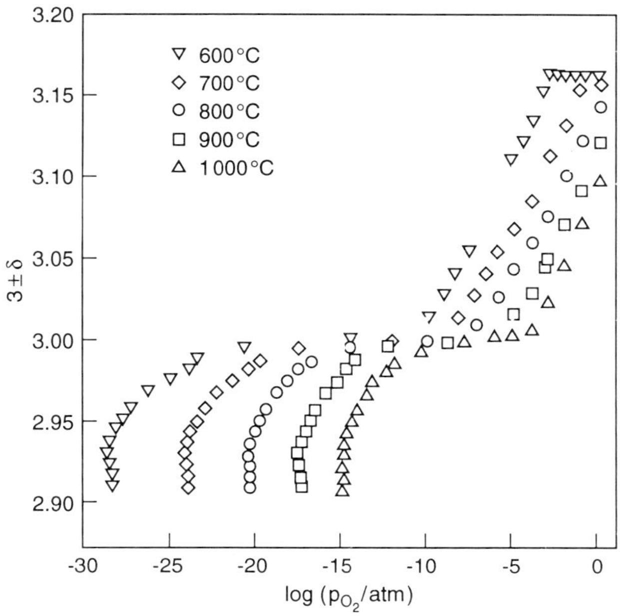

Figure 8 Nonstoichiometry of $LaMnO_3$ at 600–1000°C.

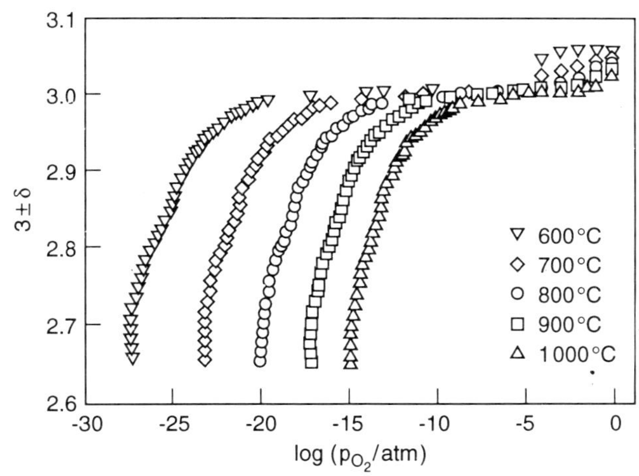

Figure 9 Nonstoichiometry of $La_{0.8}Sr_{0.2}MnO_3$ at 800°C.

agreement with those in iodometry. The relationship between oxygen content and Sr content, x, becomes nearly linear. The oxygen content for $x = 0.5$ at 600°C in 1 atm O_2, that is oxygen-to-metal ratio, gives actually equal to 3, as pointed out by Kuo *et al.*[7]

The oxygen content of the oxide in the plateau is also decided from the standpoint of perovskite-type structure. The perovskite-type structure has no interstitial sites for excess oxygen. In the hyperstoichiometric region of $La_{1-x}Sr_xMnO_{3+d}$, oxygen atoms fill the lattice points and metals at A and/or B sites are in deficiency, as shown

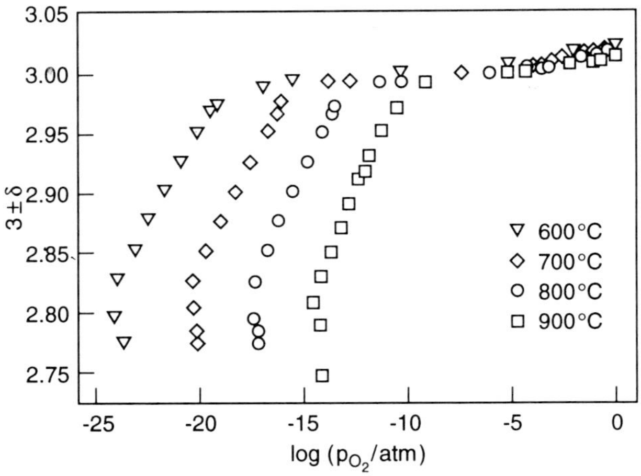

Figure 10 Nonstoichiometry of $La_{0.6}Sr_{0.4}MnO_3$ at 600–900°C.

by the formula $La_{3(1-x)/(3+d)}Sr_{3x/(3+d)}Mn_{3/(3+d)}O_3$. In the hypostoichiometric region, oxygen lattice sites are in deficiency, as shown in $La_{1-x}Sr_xMnO_{3-d}$. The existence of vacancy of metal or oxygen atoms, however, is not experimentally confirmed yet because of small amounts of vacancy, although the hyperstoichiometric and hypostoichiometric oxides have been examined in X-ray and neutron diffraction at room temperature.

3.4 Decomposition pressure

When oxygen pressure is lowered, $LaMnO_3$ is decomposed into component oxides La_2O_3, SrO and MnO. As $(La,Sr)_2MnO_4$ as a decomposition product is observed for the decomposition of $La_{1-x}Sr_xMnO_3$ ($x \geqq 0$), it is thought that $(La,Sr)MnO_3$ is first decomposed into $(La,Sr)_2MnO_4$ and MnO, then $(La,Sr)_2MnO_4$ into La_2O_3, SrO and MnO. The decomposition pressure for $La_{1-x}Sr_xMnO_3$, where $x = 0$–0.5, was measured in coulometric titration. The result is shown in Fig. 11. It is found that the decomposition pressure is nearly linearly raised with doped Sr content. The heat of decomposition slightly decreases with Sr content. The stability of $(La,Sr)MnO_3$ is shown to decrease with Sr content as a dopant.

4. CONCLUDING REMARKS

(1) The crystal structure of a perovskite-type oxide, $La_{1-x}Sr_xMnO_3$ ($x = 0, 0.1, 0.3$) was determined by high-temperature X-ray diffraction as a function of temperature and oxygen pressure and the phase relations in the system were derived.

(2) Four modifications were found in $La_{1-x}Sr_xMnO_3$ ($x = 0$–0.3): Cubic, rhombohedral, two orthorhombic structure. The structure changes from orthorhom-

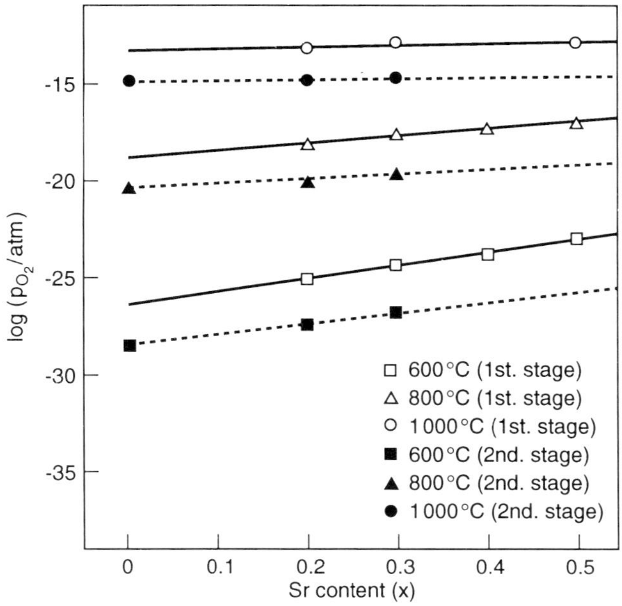

Figure 11 Decomposition pressure of (La,Sr)MnO_3 as a function of temperature and Sr content.[6]

bic to cubic through rhombohedral as Sr content increases and temperature is raised.

(3) The oxygen nonstoichiometry in $La_{1-x}Sr_xMnO_3$ ($x = 0, 0.2, 0.4$) is shown as an oxygen content vs. log p_{O_2} curve. The oxide is divided into three composition regions: hyperstoichiometric, stoichiometric and hypostoichiometric.

(4) The oxygen content in $La_{1-x}Sr_xMnO_3$ at 600°C in 1 atm O_2 was 3.16, 3.06 and 3.02 for $x = 0$, 0.2 and 0.4, respectively, and these values were in good agreement with those obtained in iodometry.

(5) At the oxygen stoichiometry in $La_{1-x}Sr_xMnO_3$ observed in reduced oxygen pressures, the oxidation state of Mn is +3 for $x = 0$ and increases with doped Sr content.

REFERENCES

1. J. Mizusaki: *Solid State Ionics*, 1992, **52**, 79–91.
2. J. Mizusaki, H. Tagawa, K. Naraya and T. Sasamoto: *Solid State Ionics*, 1991, **49**, 111.
3. F. Izumi: *J. Mineral Soc. Jpn*, 1985, **17**, 37.
4. F. S. Galasso: *Structure, Properties and Preparation of Perovskite-Type Compounds*, Pergamon Press, Oxford, 1981.
5. Y. Takeda, S. Nakai, T. Kojima, R. Kanno, N. Imanishi, G. Q. Shen and O. Yamamoto: *Mater. Res. Bull.*, 1991, **26**, 153.
6. H. Takai, H. Inaba, H. Tagawa, J. Mizusaki and T. Hashimoto: to be published.
7. J. H. Kuo, H. U. Anderson and D. M. Sparlin: *J. Solid State Chem.*, 1989, **83**, 52.

Cathodic Behaviour of Nonstoichiometric $(La,Sr)_{1-x}(Co,Mn)O_3$ Materials

J. C. C. ABRANTES,[1] C. M. S. RODRIGUES,[1]
J. A. LABRINCHA,[2] J. R. FRADE[2,*] and
F. M. B. MARQUES[2]

[1]*ESTG, Inst. Politécnico Viana do Castelo, Ap. 51, 4900 Viana do Castelo, Portugal*

[2]*Departamento de Engenharia de Cerãmica e do Vidro, Universidade de Aveiro, 3810 Aveiro, Portugal*

ABSTRACT

Solid oxide fuel cells (SOFCs) are expected to become competitive devices for electrical power generation, but successful application might be dependent on decreasing working temperatures to 800°C or lower, without detrimental effects on electrode processes and ohmic losses. This requires alternative electrode and electrolyte materials. The high electronic and oxygen ion conductivities of $(La,Sr)CoO_3$ are promising for oxygen reduction, but high temperature reaction with yttria-stabilised zirconia (YSZ) electrolytes and excessive thermal expansion mismatch have prevented the current application of these cathode materials. Expected ways to try to minimise the reactivity of cobaltates involve A-site deficiency, and partial Mn for Co substitution. The latter is also expected to contribute to lower the thermal expansion mismatch.

In this work $(La,Sr)_{1-x}(Co,Mn)O_3$ materials were prepared by solid state reaction and characterised by X-ray diffraction, thermal expansion and electrical conductivity measurements. The reactivity between these electrode materials and YSZ was also evaluated by XRD and SEM. Electrode layers were screen-printed on YSZ substrates and cathodic polarisation was measured at temperatures up to 1000°C. A-site substoichiometry and Co for Mn substitution affect the cathodic polarisation, but this also depends on some morphological features of screen-printed electrode layers.

1. INTRODUCTION

Perovskites based on $LaMnO_3$ and $LaCoO_3$, with partial replacement of La by Sr, are the most common materials which have been tested as air electrodes for solid oxide fuel cells (SOFC), working at temperatures around 1000°C. Major requirements for these materials include high electronic and oxygen–ion conductivities, thermal expansion coefficients (TEC) compatible with the electrolyte, good catalytic activity, low activation polarisation, and low reactivity towards the electrolyte or other cell components. $LaMnO_3$-based materials have TECs compatible with YSZ electrolytes[1] and low reactivity towards the same electrolyte,[2] but activation polarisation still accounts for a significant fraction of total losses,[3,4] probably because their oxygen–ion conductivity is small.[5] The electrical conductivity of

*To whom all correspondence should be sent.

$LaCoO_3$-based materials can be higher than 10^3 S/cm at 1000°C,[6] and polarisation losses might be low[7] due to a significant oxygen ion conductivity.[8] Unfortunately, these advantages could be spoiled by excessive thermal expansion mismatch between $LaCoO_3$-based materials and YSZ, and high temperature reactivity.

Partial replacement of Sr for La in $LaMO_3$ (M = Mn,Co) can cause a slight increase in electrical conductivity[2,6] and probably also improve electrocatalytic activity, which has been attributed to increase of oxygen vacancy concentration,[4] especially at moderately low oxygen partial pressures. The role of composition might thus be interpreted by considering the expected changes in defect concentrations with composition. Schottky-type defects are commonly accepted as dominant in these materials.[9] Partial replacement of Sr for La should originate the formation of electronic defects localised in M positions,[4] and cathodic overpotentials are believed to promote the formation of high concentrations of oxygen vacancies, which should be active sites for oxygen reduction. In addition, La-deficient materials seem to behave like Sr-doped compositions, having higher electrical conductivity than stoichiometric materials, and improved electrocatalytic activity.[10] Moreover, a decrease in reactivity between La-deficient perovskites and YSZ is also expected, minimising the formation of resistive zirconate phases and decreasing the ohmic losses.[11]

The aim of this work was thus the evaluation of the role of La deficiency on structure, transport properties, reactivity towards YSZ and cathodic behaviour of a group of perovskites. The use of mixed Co/Mn compositions corresponding also to an attempt to combine the best properties of both single (B-site) constituent perovskites ($LaCoO_3$ or $LaMnO_3$). Having in mind that the central objective of this work is the study of A-site deficient ABO_3 perovskites, it should be noticed that in the following text the designations 'substoichiometric' or 'nonstoichiometric' perovskites will be used to describe this A/B occupancy ratio smaller than 1. This does not exclude the possibility of other deviations from stoichiometry, namely in the metal/non-metal ratio, as commonly observed in this group of materials.

2. EXPERIMENTAL

Powders of different perovskite compositions $(La_{1-x}Sr_x)_{1-y}(Mn_{1-z}Co_z)O_3$ (with $0 \leq x \leq 0.3$; $y \leq 0.1$; $0 \leq z \leq 1$) were prepared by solid state reaction of mixtures of La_2O_3, $SrCO_3$, $MnCO_3$ and Co_3O_4, calcined at 1100°C for 10 h. Complete reaction after calcination was confirmed by X-ray diffraction. Pellets sintered in air at 1400–1500°C for 2 h were used for electrical conductivity measurements, performed according to the non-collinear four probe van der Pauw technique. Dilatometric analyses between room temperature and 1000°C were performed using samples cut from sintered pellets. Electrode layers were deposited on YSZ substrates by screen-printing and fired at 1250°C for 30 min.

Polarisation measurements were performed up to 1000°C with a three electrodes technique, using porous platinum reference and counter electrodes. The ohmic components between reference and working electrodes were evaluated by im-

Table 1
Structural characterisation of perovskites.

	Rhombohedric		Hexagonal		$10^7 \times$ TEC
Composition	a_o	α	A_o	C_o	(°C^{-1})
$(La_{0.7}Sr_{0.3})_{0.9}CoO_{3-\delta}$	3.832	89.94	5.416	13.29	142
$La_{0.7}Sr_{0.3}CoO_{3-\delta}$	3.838	90.06	5.433	13.28	148
$(La_{0.7}Sr_{0.3})_{0.9}Co_{0.7}Mn_{0.3}O_{3-\delta}$	3.846	90.08	5.442	13.30	134
$La_{0.7}Sr_{0.3}Co_{0.7}Mn_{0.3}O_{3-\delta}$	3.850	90.02	5.445	13.33	146
$(La_{0.7}Sr_{0.3})_{0.9}Co_{0.3}Mn_{0.7}O_{3-\delta}$	3.838	90.09	5.431	13.27	114
$La_{0.7}Sr_{0.3}Co_{0.3}Mn_{0.7}O_{3-\delta}$	3.864	89.98	5.464	13.39	123

pedance spectroscopy. Results on electrode/electrolyte reactivity were obtained after thermal treatment of powder mixtures (YSZ and cathode materials) at temperatures between 1100 and 1400°C during 2–4 days. The electrolyte powder was YSZ with 8 mol% Y_2O_3 (Tosoh Co.) and the amount of reaction products formed was evaluated by XRD using the internal standard method.[6] XRD, SEM and EDS analyses were used to characterise the materials structure and electrode layer microstructures.

3. RESULTS AND DISCUSSION

XRD results of calcined samples show complete reaction of starting reagents and formation of a perovskite-type phase indexed to the hexagonal–rhombohedric system. No second phase could be noticed, which suggests that composition excursions (namely for highly defective substoichiometric compositions) could be accommodated within the base structure. The lattice parameters of Co-rich compositions are lower than those of manganites and tend to decrease with La-deficiency (Table 1). A-site deficiency also enhances the difference $|\alpha - 90°|$, which corresponds to deviations from cubic symmetry. The effects of A-site deficiency are probably related to an increase of the $[M^{4+}]/[M^{3+}]$ ratio and/or an increase of oxygen-vacancy concentration, required for charge compensation.

Table 1 also shows the changes of TECs as a function of A-site deficiency, and Mn content in Co/Mn mixed compositions. According to common knowledge the values of TEC decrease as Mn content increases, and compositions with high Mn contents are compatible to the thermal expansion of YSZ ($\approx 100 \times 10^{-7}$ °C^{-1}). However, A-site deficiency also contributes to decrease the TEC of all compositions tested, improving the compatibility towards the substrate.

Results of electrical conductivity measurements shown in Fig. 1 indicate that the behaviour of Mn-rich materials is of a semiconductor type, and Co-rich compositions are of a metallic type, even at room temperatures, with conductivities much higher than Mn-rich compositions. Differences in high temperature conductivity

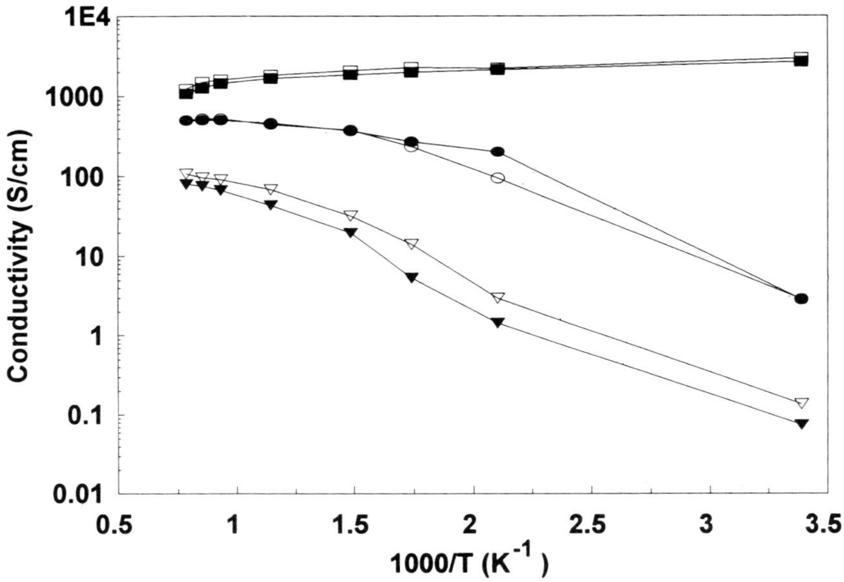

Figure 1 Electrical conductivity of cathode materials. □, $(La_{0.7}Sr_{0.3})_{0.9}CoO_{3-\delta}$; ■, $(La_{0.7}Sr_{0.3})CoO_{3-\delta}$; ○, $(La_{0.7}Sr_{0.3})_{0.9}Co_{0.7}Mn_{0.3}O_{3-\delta}$; ●, $(La_{0.7}Sr_{0.3})Co_{0.7}Mn_{0.3}O_{3-\delta}$; ▽, $La_{0.9}MnO_{3-\delta}$; ▼, $LaMnO_{3\pm\delta}$.

Table 2
Reaction products formed by reaction between YSZ and electrode materials, during 48 h.

Composition	Temperature (°C)	$\%SrZrO_3$	$\%La_2Zr_2O_7$
$La_{0.7}Sr_{0.3}CoO_{3-\delta}$	1100	9	36
	1200	12	52
$(La_{0.7}Sr_{0.3})_{0.9}CoO_{3-\delta}$	1200	13	48
$La_{0.7}Sr_{0.3}Mn_{0.7}O_{3-\delta}$	1100	0	0

between Mn-rich and Co-rich compositions are about one order of magnitude.[2,6] A-site deficiency tends to slightly increase the electrical conductivity values in air (Fig. 1), as previously reported for La-deficiency in Sr-free samples.[12]

The reactivity between electrode materials and YSZ at different temperatures is given in Table 2; 30% Sr doped samples were selected based on a previous finding that formation of reaction products is minimal for this Sr content.[6] Note that increasing Sr contents promotes the formation of strontium zirconate and hinders the formation of lanthanum zirconate, and the electrical resistivities of both reaction products are 2–3 orders of magnitude higher than YSZ.[11,14] For the limited range of compositions now exploited, cobaltates were found much more reactive than the manganites, as previously reported.[2,6] A-site deficiency should also contribute to

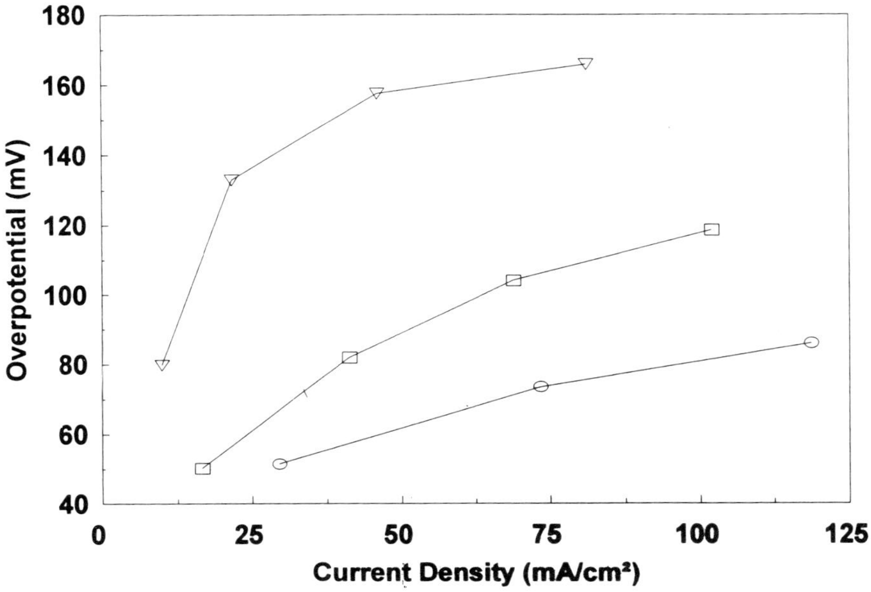

Figure 2 Cathodic polarisation curves obtained with the following A-site deficient cathodes: ○, $(La_{0.7}Sr_{0.3})_{0.9}Co_{0.7}Mn_{0.3}O_{3-\delta}$; □, $(La_{0.7}Sr_{0.3})_{0.9}Co_{0.3}Mn_{0.7}O_{3-\delta}$; ▽, $(La_{0.7}Sr_{0.3})_{0.9}CoO_{3-\delta}$.

lower reactivity, as found for $La_{1-x}MnO_{3-\delta}$,[10] but the results obtained for cobaltate-based materials are quite identical for both stoichiometric and sub-stoichiometric materials (Table 2).

The effects of Co/Mn ratio and A-site deficiency on cathodic polarisation of screen-printed layers were also evaluated (Fig. 2). A-site deficiency is believed to decrease the cathodic overpotential at a constant current density, because of the role of oxygen vacancies in the electrode process. Oxygen vacancies might be present in these materials for charge compensation of cation vacancies. For high cathodic overpotentials one expects a decrease of oxygen partial pressure near the triple contact points, which can also favour an increase in oxygen ion conductivity.[4] Note that small differences in electrical conductivity between stoichiometric and A-site deficient materials (Fig. 1) are unlikely to cause a significant change on electrode polarisation behaviour, but might be relevant in terms of determining the concentration of active sites for the electrode reaction.

Figure 2 shows that the cathodic polarisation required to attain a given current density is lower for manganese-containing samples than for cobaltates, in contradiction to the expected behaviour of this last group of materials. The poor electrode performance of screen-printed cobaltate layers is probably due to the reactivity of cobaltates (Table 2) at firing temperatures, or even after long times at working temperatures in the range 1000–1100°C. For example, Fig. 3 shows thin layers of reaction products formed at the interface between the Mn-free cathode composition and the YSZ electrolyte, while the remaining interfaces appear free of such reaction

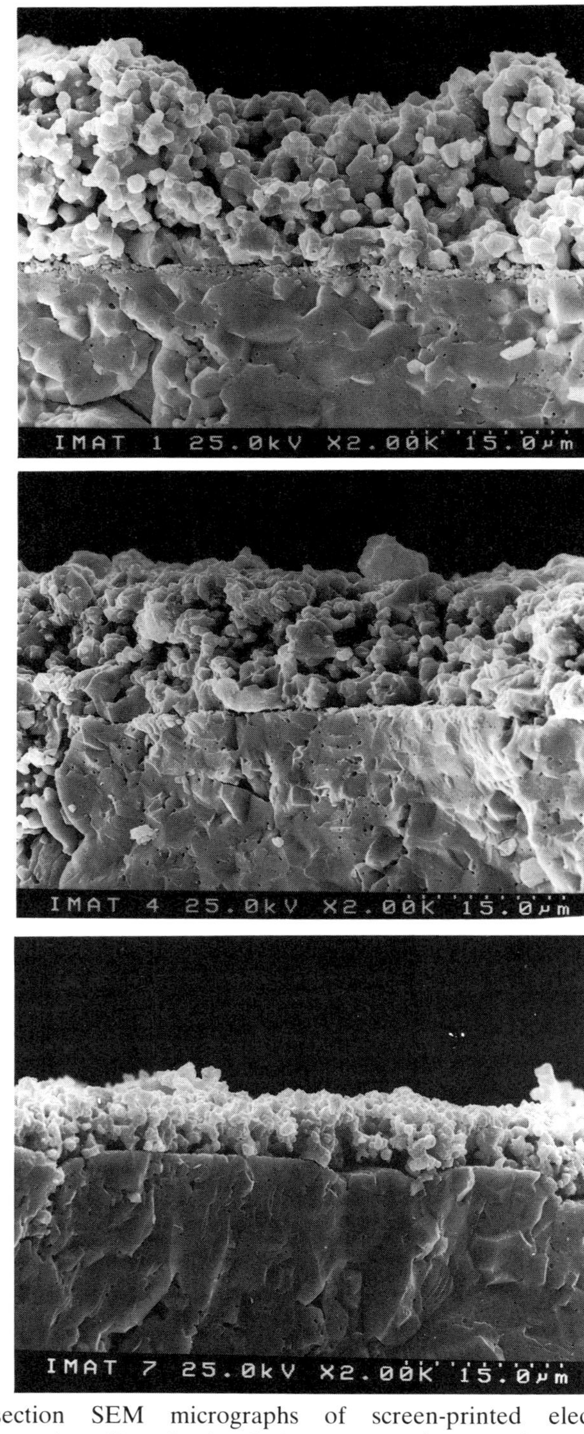

Figure 3 Cross-section SEM micrographs of screen-printed electrode layers on YSZ substrates. A, $(La_{0.7}Sr_{0.3})_{0.9}CoO_{3-\delta}$; B, $(La_{0.7}Sr_{0.3})_{0.9}Co_{0.7}Mn_{0.3}O_{3-\delta}$; C, $(La_{0.7}Sr_{0.3})_{0.9}Co_{0.3}Mn_{0.7}O_{3-\delta}$.

products. These SEM observations of layers fired under identical conditions, at 1250°C, also agree with reported qualitative trends in electrode/electrolyte reactivity.

As mentioned, Fig. 2 shows changes of cathodic overpotentials on varying the Mn : Co ratio of $(La_{0.7}Sr_{0.3})_{0.9}Co_yMn_{1-y}O_3$ materials. A relatively low polarisation is achieved for the 70Co–30Mn mixed composition, probably arising from a compromise between good intrinsic electrochemical performance expected for Co-rich compositions, and suitable microstructures without significative formation of reaction products. The 30Co–70Mn electrode has an intermediate behaviour, in spite of the absence of reaction products, and what appears to be a relatively good microstructure. Electrode deposition experiments have also shown that A-site deficiency in Co-containing cathodes lowers the risk of poor adhesion or failure by excessive thermal expansion mismatch between YSZ substrates and electrode layers.

4. PROPOSED DEFECT CHEMISTRY FOR A-SITE DEFICIENT MATERIALS

The behaviour of substoichiometric materials and the improved performance of (at least some) these electrodes suggest that the advantages of these materials may be related to improved compatibility towards the electrolyte and better ionic transport properties, or an enhanced role of oxygen vacancies in the electrode process. Note that enhancing the ionic transport properties does not imply any significant increase in electronic or total conductivity, especially for applications where the performance of materials may be critically dependent on a minority carrier such as oxygen vacancies.

Experimental data and methods capable of detecting changes in concentrations or of the nature of dominant charge carrying defects, related to A-site deficiency, can be found in the literature. The reported drop in lattice parameters of substoichiometric compositions indicates that changes in composition correspond to increasing concentrations of vacancies. Weight losses (and the average valence of Mn or Co) as a function of oxygen partial pressure have also provided a sound basis for defect chemistry models.[9] Data reported by Mizusaki[15] show that significant changes in average Mn valence occur for values of oxygen partial pressure higher than about 1 Pa in $La_{1-x}MnO_{3-\delta}$ (at 900°C), and $Po_2 < 10^{-15}$ Pa for $LaMnO_{3\pm\delta}$ or $La_{1-y}Sr_yMnO_{3-\delta}$. Charge compensation of negatively charged defects (Sr'_{La} or V'''_{La}) should be provided by oxygen vacancies when the average valence of Mn (or Co) is 3, and by electron holes for higher values of average valence.

The following equations correspond to using the basic model proposed by Anderson and coauthors,[9] with the Kroger–Vink notation for defects:

$$0 \rightleftarrows V'''_A + V'''_B + 3V^{\cdot\cdot}_O \tag{1}$$

$$K_S = [V'''_A][V'''_B][V^{\cdot\cdot}_O]^3 \tag{2}$$

$$2B_B \rightleftarrows B_B' + B_B^{\cdot} \tag{3}$$

$$K_1 = [B_B'][B_B^{\cdot}]/[B_B]^2 \tag{4}$$

$$O_O + 2B_B^{\cdot} \rightleftarrows 2B_B + V_O^{\cdot\cdot} + 1/2O_2 \tag{5}$$

$$K_2 = ([B_B]/[B_B^{\cdot}])^2[V_O^{\cdot\cdot}]Po_2^{1/2} \tag{6}$$

with K_S, K_1 and K_2 being the equilibrium constants for these reactions. The electroneutrality condition provides an additional relation between concentrations of charged defects:

$$[B_B^{\cdot}] + 2[V_O^{\cdot\cdot}] = y + [B_B'] + 3[V_A'''] + 3[V_B'''] \tag{7a}$$

with $y = [Sr_{La}']$ representing the strontium content. The concentration of B-site positions must also be constant:

$$[B_B] + [B_B^{\cdot}] + [B_B'] + [V_B'''] = b. \tag{8}$$

The last relation between concentrations is related to substoichiometry, which changes the relation between cationic vacancies predicted for Schottky-type disorder; this yields:

$$[V_A'''] = [V_B'''] + x. \tag{9}$$

Equations (2), (4) and (6–9) can be solved for given values of x (substoichiometry), Sr content (y), and equilibrium constants K_S, K_1 and K_2, to obtain the expected effects of oxygen partial pressure on defect concentrations. The total concentrations of A- or B-site positions can be evaluated from the relevant unit cell volume. Actually, experimental data suggest that, in these materials, the contribution of intrinsic Schottky disorder is usually negligible, and this can be used to simplify the electroneutrality condition.[15] As formation of cation vacancies is still predicted for A-site deficient materials, $[V_A'''] \approx x$, and in this case eq. (7a) reduces to:

$$[B_B^{\cdot}] + 2[V_O^{\cdot\cdot}] \approx y + [B_B'] + 3x. \tag{7b}$$

The remaining equilibrium constants can be adjusted to yield the expected ranges of oxygen partial pressures where the average valence of B is significantly different from 3 (see Mizusaki[15]).

Typical examples of these calculations are shown in Fig. 4, to match the experimental changes in Mn valence, or weight gain in $La_{1-y}MnO_{3-\delta}$ and $LaMnO_{3\pm\delta}$ materials.[15,16] From this figure one expects that the transition $V_O^{\cdot\cdot} \rightarrow h^{\cdot}$ must be displaced towards higher oxygen partial pressures in A-site deficient materials than for the case of Sr-containing samples. The proposed defect diagram

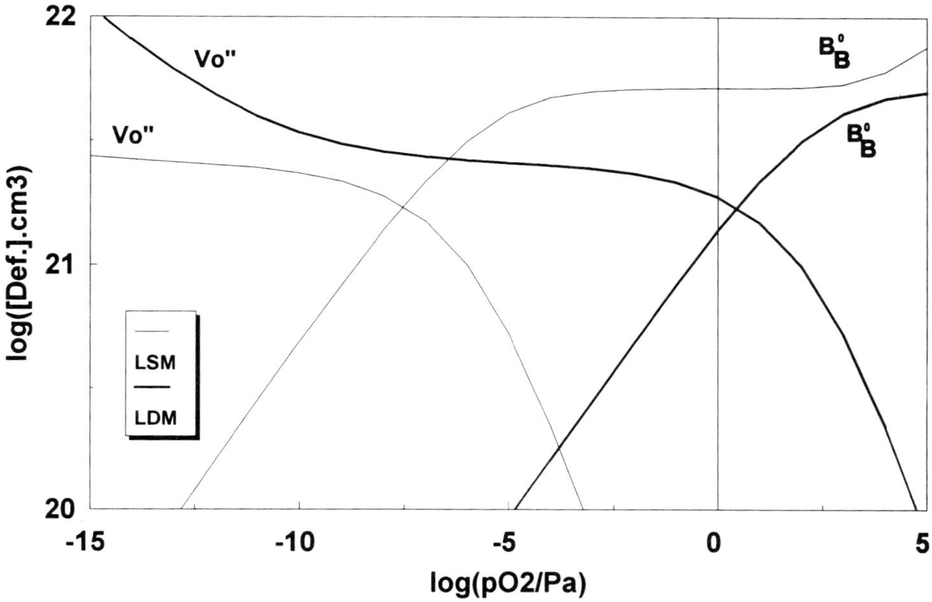

Figure 4 Partial defect diagrams proposed for $La_{0.7}Sr_{0.3}MnO_{3\pm\delta}$ (LSM), and $La_{0.9}MnO_{3-\delta}$ (LDM) showing the expected ranges of values of oxygen partial pressures for the $V_o^{\cdot\cdot} \rightarrow h^{\cdot}$ transition.

also suggests that A-site deficiency might be effective in originating significant concentrations of oxygen ion vacancies and ionic conductivity in relatively oxidising conditions, which is unlikely to be achieved by Sr for La substitution. The ionic contribution of Co-rich compositions in air is also relatively high,[17] but these materials are less stable than the manganites and decompose at high oxygen partial pressures.[13]

This discussion suggests that A-site deficient materials still deserve further attention. To obtain a complete evaluation of relevant parameters in electrode performance more information is required aiming at an independent evaluation of oxygen ion conductivity, electrocatalytic efficiency for oxygen reduction (with point type electrodes or physically deposited layers with controlled microstructures), and/or oxygen exchange behaviour. The performance of electrode layers fired at moderate temperatures was indeed found sensitive to the combined effect of different parameters, and partly inconclusive for this reason.

4. CONCLUSIONS

Substoichiometry and Mn for Co substitution might improve the compatibility between YSZ electrolytes and generic electrode materials $La_{1-x-y}Sr_yCo_{1-z}Mn_zO_{3-\delta}$. Both changes contribute to minimise the limitations of

Co-rich compositions, related to formation of unwanted reaction products, and thermal expansion mismatch. However, Mn for Co substitution might spoil the excellent electrochemical performance of $LaCoO_3$ based cathode materials, which are efficient catalysts for oxygen reduction. In fact, the concentration of oxygen vacancies of Mn-rich compositions is probably low in oxidising conditions. On the contrary, A-site deficiency is expected to enhance the concentration of oxygen vacancies. Lower cathodic polarisations observed when using these electrode materials might be explained by this compositional effect. Because of the combined role of different parameters, the experimental evidence can be considered inconclusive in terms of the effective role of substoichiometry in electrode performance.

ACKNOWLEDGEMENTS

This work was funded by JNICT (Portugal) under Contract SRTDB/CTM/C/664/92.

REFERENCES

1. A. Hammouche, E. Siebert and A. Hammou: *Mat. Res. Bull.*, 1989, **24**(3), 367.
2. J. A. Labrincha, J. R. Frade and F. M. B. Marques: *Silicates Industriels*, 1994, **LIX**, 59.
3. Y. Takeda, R. Kanno, M. Noda, Y. Tomida and O. Yamamoto: *J. Electrochem. Soc.*, 1987, **134**, 2656.
4. A. Hammouche, E. Siebert, A. Hammou, M. Kleitz and A. Caneiro: *J. Electrochem. Soc.*, 1991, **138**, 1212.
5. B. C. H. Steele, S. Carter, J. Kajda, I. Kontoulis and J. A. Kilner: 'Optimisation of fuel cell components using $^{18}O/^{16}O$ exchange and dynamic SIMS techniques', in *Proc. 2nd Int. Symp. Solid Oxide Fuel Cells*, F. Gross, P. Zegers, S. C. Singhal and O. Yamamoto (eds), Commission European Communities, Luxembourg, 1991, pp. 517–525.
6. J. A. Labrincha, J. R. Frade and F. M. B. Marques: 'Reaction between cobaltate cathodes and YSZ', in *Proc. 2nd Int. Symp. Solid Oxide Fuel Cells*, F. Gross, P. Zegers, S. C. Singhal and O. Yamamoto (eds), Commission European Communities, Luxembourg, 1991, pp. 689–96.
7. O. Yamamoto, Y. Takeda, R. Kanno and M. Noda: *Solid State Ionics*, 1987, **22**, 241.
8. B. C. H. Steele, 'Ceramic materials for SOFC and electrochemical reactor technology', in *Electroceramics IV*, vol. II, R. Waser *et al.* (eds), Augustinus Buchhandlung, Germany, 1994, pp. 663–676.
9. H. U. Anderson, J. H. Kuo and D. M. Sparlin: 'Review of defect chemistry of $LaMnO_3$ and $LaCrO_3$', in *Proc. First Int. Symp. Solid Oxide Fuel Cells*, S. C. Singhal (ed.), *The Electrochem. Soc. Proc.*, 1989, vol. 89–11, pp. 68–77.
10. J. C. C. Abrantes, F. M. B. Marques and J. R. Frade: 'Electrical conductivity of $La_{1-x}MnO_3$ perovskites', in *Euroceramics II*, 1991, vol. 3, p. 2167, G. Ziegler *et al.* (eds), Deutsche Keramische Gesellschaft, Koln.
11. J. A. Labrincha, J. R. Frade and F. M. B. Marques: *J. Mat. Sci.*, 1993, **28**, 3809.
12. Y. Takeda, S. Nakai, T. Kojima, R. Kanno, N. Imanishi, G. Q. Shen, O. Yamamoto, M. Mori, C. Asakawa and T. Abe: *Mat. Res. Bull.*, 1991, **26**, 153.
13. J. Mizusaki, H. Tagawa, K. Naraya and T. Sasamoto: *Solid State Ionics*, 1991, **49**, 111.

14. J. A. Labrincha, J. R. Frade and F. M. B. Marques: *Solid State Ionics*, 1993, **61**, 71.
15. J. Mizusaki: *Solid State Ionics*, 1992, **52**, 79.
16. H. U. Anderson: *Solid State Ionics*, 1992, **52**, 33.
17. Y. Teraoka, H. M. Zhang, K. Okamoto and N. Yamazoe: *Mat. Res. Bull.*, 1988, **23**, 51.

Cathodic Behaviour of ZrO_2-doped In_2O_3 on 8 mol% Y_2O_3-stabilised ZrO_2 Electrolyte for Solid Oxide Fuel Cell

J.-M. BAE, H. FOX, J. A. KILNER and B. C. H. STEELE
Centre for Technical Ceramics, Department of Materials, Imperial College, London SW7 2BP, UK

ABSTRACT
Thick film (≈50 μm) of $InO_{1.5}$–1 mol% ZrO_2 (IZO-1) was fabricated on both sides of 8 mol% Y_2O_3-stabilised ZrO_2 (8YSZ). AC impedance spectroscopic measurements were carried out to investigate the electrodic behaviour of IZO-1 on 8YSZ with function of temperature, oxygen partial pressure (Po_2) and the composition of IZO. The AC responses mainly consist of three semicircles. Thermal expansion coefficient of IZO-1 was measured from room temperature to 1000°C and compared with that of 8YSZ. The reactivity of IZO-1 with 8YSZ was checked by X-ray diffraction. Oxygen self-diffusion coefficient and surface exchange coefficient at IZO-1 were measured by $^{18}O/^{16}O$ isotope exchange and following measurement of the profile using the secondary ion mass spectroscopy (SIMS) technique.

1. INTRODUCTION

Power loss of solid oxide fuel cell (SOFC) can be effectively reduced by using an appropriate cathode.[1] Cathodic materials should have pertinent thermal expansion coefficient to electrolyte and interconnector and be thermodynamically stable at cathodic (oxidising) condition. A cathode should be chemically inert with other components, i.e. electrolyte and interconnector. Cathodic materials are generally high electrical conductors. Mixed conduction of oxygen ion and electron/hole is known as a contribution for the decrease of cathodic polarisation.[2,3]

ZrO_2-doped In_2O_3 (IZO) has high electrical conductivity.[4] Thermal expansion coefficient of this material can be tailored by controlling the amount of ZrO_2. IZO does not make high resistive materials with Y_2O_3-stabilised ZrO_2 (YSZ) electrolyte.

In this study IZO thick film is made on YSZ and the electrodic characteristics are investigated at various temperatures and partial oxygen pressures by the impedance spectroscopy method. The diffusion and surface exchange coefficients of oxygen in IZO were measured by the secondary ion mass spectroscopy (SIMS) method.

2. EXPERIMENTS

2.1 *Isotopic $^{16}O/^{18}O$ Exchange-diffusion Experiment and Dilatometry*

The $InO_{1.5}$-1 mol% ZrO_2 (IZO-1) powders were prepared by the well-defined coprecipitation method. The starting materials were indium nitrate pentahydrate

99.99% and zirconyl chloride octahydrate 98% of Aldrich Co. After weighing both source materials as 99 : 1 of the molar ratio of cation, i.e. In : Zr, they were dissolved into the de-ionised water. The solution was diluted to 0.1 mol/l. The solution was dropped very slowly into an ammonium solution with stirring. The precipitated gel was obtained by centrifuge. To remove the water in the gel completely, the gel was boiled for 6 h in heptanol (boiling point: 175°C). After drying these powders for 48 h, ball milling in ethanol was carried out for 24 h. It was calcined at 1100°C for 2 h after drying out the ethanol.

13 mm circular-type pellets of IZO-1 were uniaxially pressed with 15.1 MPa pressure. The green bodies were sintered at 1500°C for 10 h. The IZO-1 pellets were polished into 0.25 μm roughness for $^{18}O/^{16}O$ isotope exchange. They were preannealed in 1 atm of dry ^{16}O at the same temperature as following ^{18}O anneal but for a longer time by order of magnitude to ensure equilibrium of ^{16}O concentration in the samples. ^{18}O isotope anneals were carried out for 500 s at 513, 795, and 982°C in 97% of ^{18}O at 1 atm. To obtain the depth profile of ^{18}O by SIMS (Atomika 6500), the exchanged samples were cut perpendicular to the polished surface and the new surfaces were polished again into 0.25 μm roughness. The depth profiles were fitted to the following equation[5] which is the solution of Fick's second law for the semi-infinite medium:

$$\frac{C_x - C_{bg}}{C_g - C_{bg}} = \text{erfc}\left[\frac{x}{2\sqrt{D^* t}}\right] - [\exp(hx + h^2 D^* t)] \cdot \text{erfc}\left[\frac{x}{2\sqrt{D^* t}} + h\sqrt{D^* t}\right] \quad (1)$$

where:

C_x = the concentration of ^{18}O at the distance x
C_{bg} = the background concentration of the natural ^{18}O
t = time of ^{18}O anneal
D^* = O^{-2} self-diffusion coefficient (cm^2/s)
k = surface exchange coefficient (cm/s)
C_g = the concentration of ^{18}O anneal gas
$h = \dfrac{k}{D^*}$

having the boundary condition as follows:

$$-D^* \frac{dC}{dx} = k(C_g - C_s), \text{ at } x = 0 \text{ (solid surface)} \quad (2)$$

C_s = the concentration of ^{18}O at solid surface ($x = 0$)

which means the oxygen flux across the solid surface is proportional to the difference of concentration in gas and the solid surface.

For the dilatometry sintered body, IZO-1 was cut into a bar shape with a diamond saw. An 8 mol% Y_2O_3-ZrO_2 (8YSZ) sample was prepared to compare the thermal

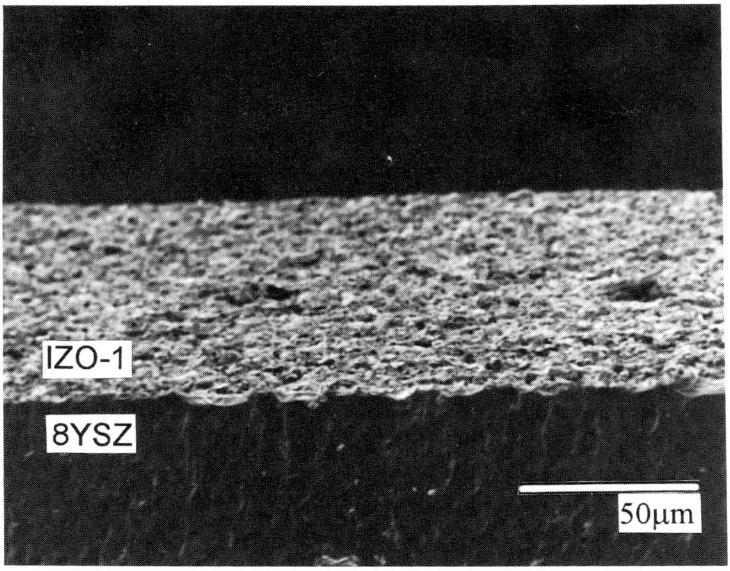

Figure 1 Microstructure of fractured surface of an IZO-1/8YSZ symmetric sample.

expansion coefficients at the same dilatometer. 8YSZ powder was packed with 15.1 MPa and sintered at 1500°C for 5 h. The thermal expansion coefficients, α, were measured from room temperature to 1000°C using a constant ramping ratio of 5°C per minute.

2.2 *Impedance Spectroscopy IZO-1/8YSZ/IZO-1 Symmetric Sample and the Reactivity of IZO-1 and 8YSZ*

Ethyl cellulose and di-*n*-phthalate were mixed by heating to 160°C with a weight ratio 1:14. IZO-1 paste was made by mixing this slurry and the IZO-1 powders with an agate pestle and mortar. This paste was brushed on both sides of 8YSZ green pellets. These pellets were co-fired at 1450°C for 4 h after drying. The fractured surface of this sample was observed by a Rigaku T220 Scanning Electron Microscope (SEM) as shown in Fig. 1. To check the reactivity of IZO-1 and 8YSZ, two powders were thoroughly mixed with the same cationic ratio and uniaxially pressed. This pellet was sintered at 1450°C for 24 h. The pellet was ground and scanned by a Philips PW1710 X-ray diffractometer with Cu-K_α radiation. Complex impedance of the symmetric samples was measured with a computer-aided Schlumberger SI 1260 Impedance/Gain-Phase Analyser at the frequency range of 0.001 to 30 MHz with changing temperatures and oxygen partial pressures. The oxygen partial pressure was controlled by mixing oxygen and argon gas and measured with a zirconia cell.

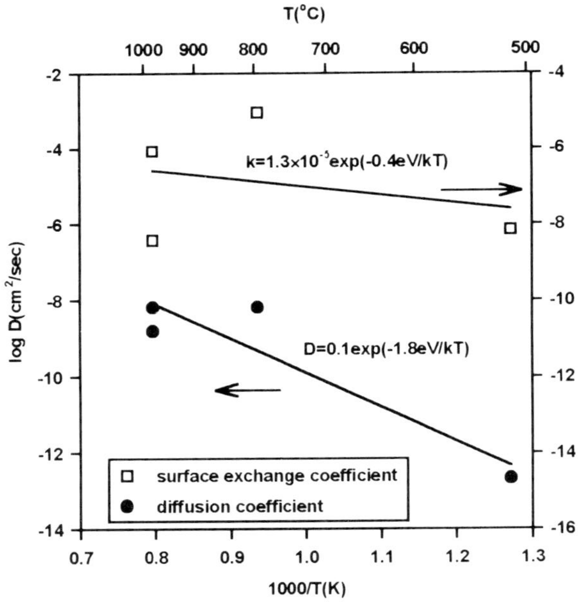

Figure 2 Diffusion and surface exchange coefficient of oxygen in IZO-1.

Table 1
Thermal expansion coefficients of IZO-1 and 8YSZ

Material	Thermal expansion coefficient (/K)
IZO-1	8.8×10^{-6}
8YSZ	9.7×10^{-6}

3. RESULTS AND DISCUSSIONS

Figure 2 is the Arrhenius plots of the diffusion and surface exchange coefficients of oxygen atom in IZO-1. Wirtz *et al.*[6] reported diffusion coefficients of In_2O_3 as $D = 0.029\ \exp(-1.8\,\text{eV/kT})$ [cm²/s], and Berger *et al.*[7] as $0.102\ \exp(-1.6\,\text{eV/kT})$ [cm/s] for Sn doped In_2O_3 (ITO). Diffusion and surface exchange coefficients of IZO-1 in this work are $D = 0.1\exp(-1.8\,\text{eV/kT})$ [cm²/s] and $k = 1.3 \times 10^{-5}$ $\exp(-0.4\,\text{eV/kT})$ [cm/s]. The comparative Arrhenius graphs are shown in Fig. 3.

The linear thermal expansion coefficients (α) of IZO-1 and 8YSZ are given in Table 1. This shows mean linear thermal expansion coefficients of IZO and 8YSZ at the range of room temperature to 1000°C. The α value of IZO-1 is smaller than that of 8YSZ so that the IZO-1 layer is under compressive stress during the cooling process of fabrication and tensile stress during heating up to operation temperature (700–900°C).

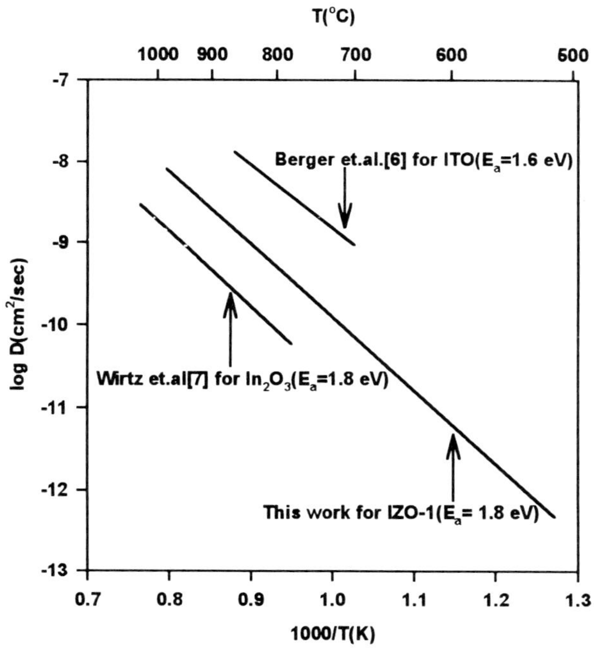

Figure 3 Diffusion coefficients of oxygen in In_2O_3, IZO-1 and ITO.

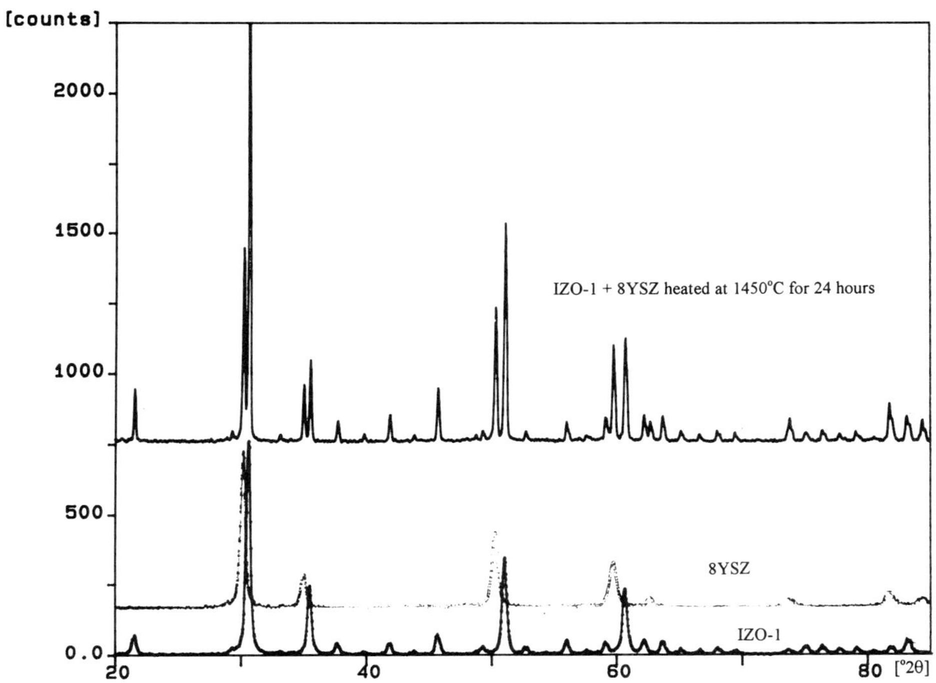

Figure 4 XRD peaks of IZO-1, 8YSZ, and IZO-1 + 8YSZ treated at 1450°C for 24 h.

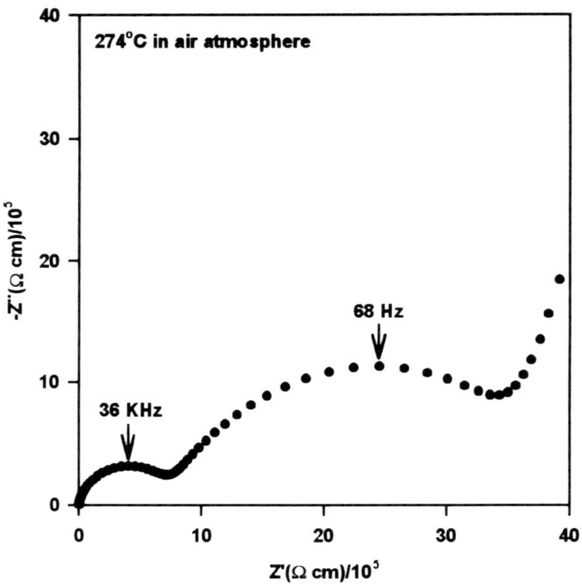

Figure 5 Complex impedance of 8YSZ electrolyte.

To check the reactivity of IZO-1 with 8YSZ, two powders were mixed by total cationic ratio, 1 : 1, and uniaxially pressed. The pellet was sintered at 1450°C for 24 h. This pellet was ground and the phase was checked by XRD. Figure 4 shows the XRD result of a mixture of IZO-1 and 8YSZ. There are no other significant peaks except those of IZO-1 and 8YSZ as seen. Therefore, there is not a significant reaction between them in the accuracy of XRD.

At complex impedance measurements the effects of 8YSZ electrolyte were observed for the high frequency region (>10 Hz) under 500°C. Two semicircles of 8YSZ can be found in Fig. 5. Resistivity of 8YSZ is $\rho = 1.3 \times 10^{-5} \exp(1.2\,\mathrm{eV/kT})$ [Ω cm] as seen in Fig. 6. This result can be extrapolated to reported data.[8,9] Over 500°C the responses of 8YSZ were so small that only responses of electrodes could be seen. Complex impedances of electrode were obtained at various temperatures and oxygen partial pressures (P_{O_2}). There are three semicircles in complex impedance of IZO-1 electrode as seen in Fig. 7. Figure 8 shows dependence of resistivities on oxygen partial pressure. The satellite in the region of high frequency has little dependence of P_{O_2}. However, those in the middle and low frequency range have strong P_{O_2} dependence. Over 10^{-2} atm of P_{O_2} resistivity of low frequency region has $\sim P_{O_2}^{-1}$ dependence, at middle frequency $\sim P_{O_2}^{-0.5}$ dependence. The characteristic relaxation time constants, τ(s), the reciprocal value of the frequency (ω_m) at the top of a satellite, are given with the function of P_{O_2} in Fig. 9. Whereas τ of high and low frequency region have little dependence on P_{O_2}, that of middle frequency has $\sim P_{O_2}^{-0.5}$ dependence. The capacity which is the representative value

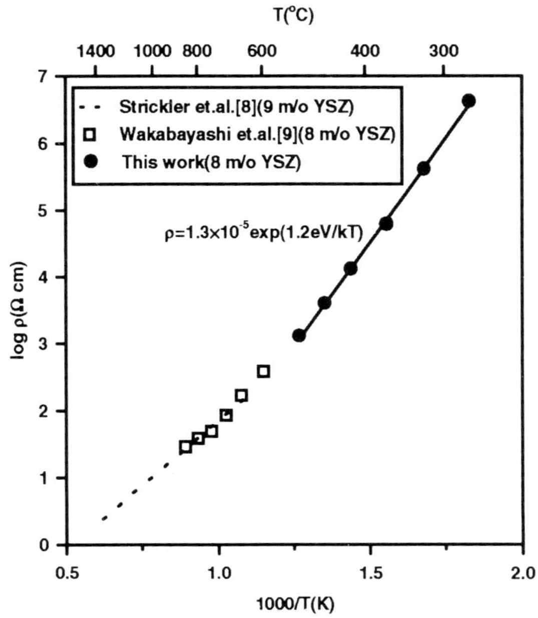

Figure 6 Resistivities of YSZ electrolytes.

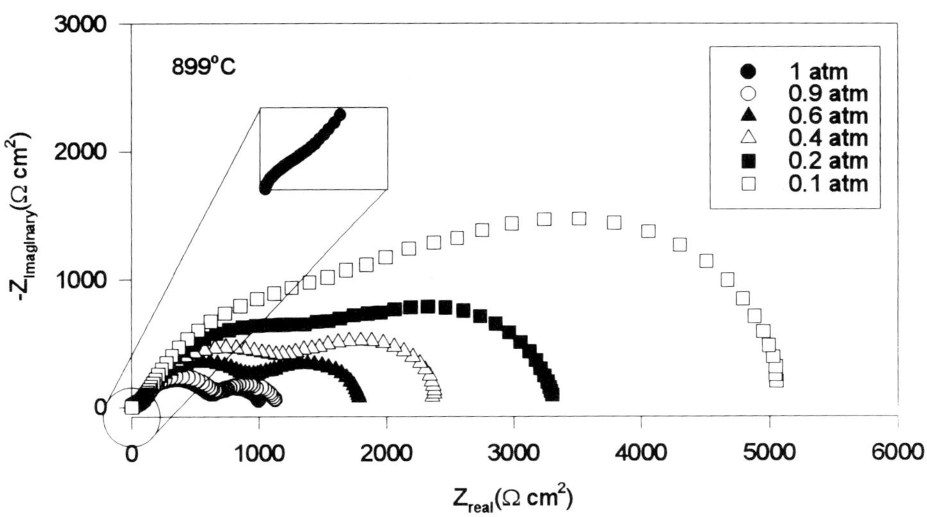

Figure 7 Complex impedances of IZO-1/8YSZ at 899°C.

an electrode reaction can be defined by:

$$C = \frac{\tau}{\rho} (\mathrm{F/cm^2}) \tag{3}$$

where ρ is electrode resistivity ($\Omega\ \mathrm{cm}^2$).

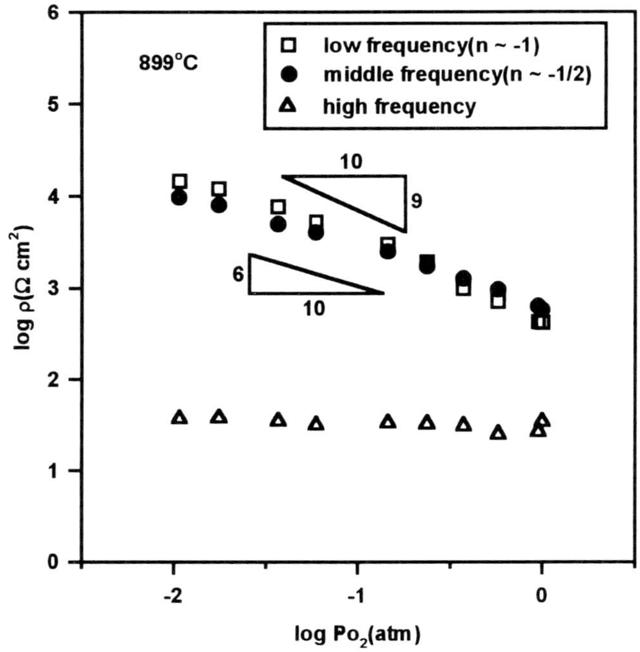

Figure 8 P_{O_2} dependence of resistivity.

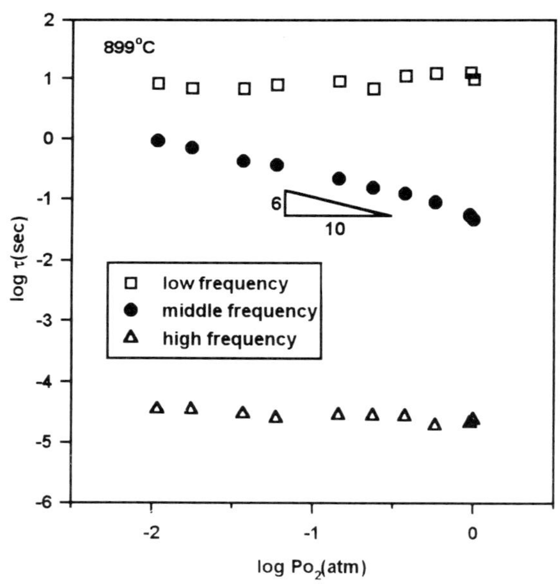

Figure 9 P_{O_2} dependence of characteristic relaxation time constant.

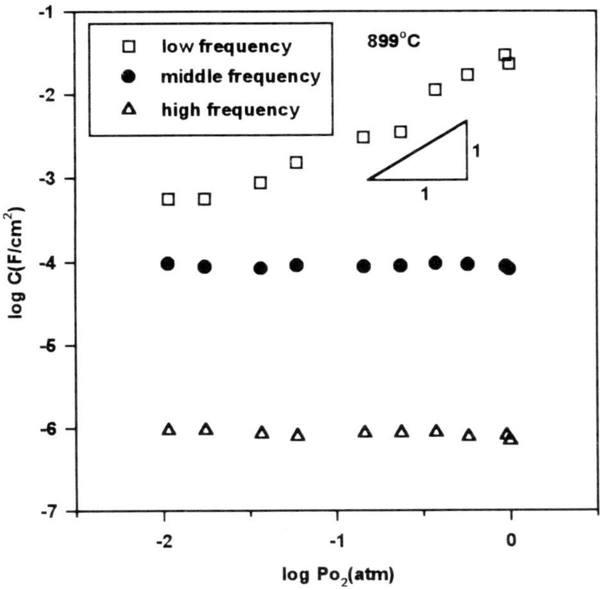

Figure 10 P_{O_2} dependence of capacity.

Each capacity at the frequency region can be calculated by this equation. Figure 10 shows the dependence of capacity at each frequency region on P_{O_2}. The capacity of low frequency range shows $P_{O_2}^{+1}$ behaviour. The capacity of the middle frequency range is almost independent of P_{O_2}.

There are several steps of reaction occurring for cathodic reaction such as gas phase diffusion of oxygen, adsorption and dissociation of oxygen, diffusion of oxygen to the interface of electrode/electrolyte in electrode, and charge transfer reaction. At low frequency range, the resistivity and the capacity have $P_{O_2}^{+1}$ dependence. This dependence can claim the semicircle at the low frequency region and is related to adsorption of oxygen oxygen molecule:

$$O_{2_{gas}} \leftrightarrow O_{2_{ads}}$$

and/or gas phase diffusion of oxygen molecule. By calculations based on a measured voltage amplitude of AC impedance about 10 mV and the capacity at low frequency range, 5.6×10^{14} atoms/cm^2 is obtained at the impedance data of 1 atm. This value means an oxygen atom can be found every 17 Å^2. If we assume the oxygen molecules at gas phase are uniformly distributed, then about 7×10^{12} atoms/cm^2 exist at 899°C and 1 atm on the surface of IZO-1, 100 times smaller than the value from the capacity. Therefore, the satellite at low frequency range may involve not only gas phase diffusion but adsorption reaction of the oxygen molecule. The characteristic relaxation time constant of this reaction is about 10 s, independent of oxygen partial pressure. The resistivity, characteristic relaxation time, and capacity of the satellite at high frequency range have little P_{O_2} dependence as shown in Figs

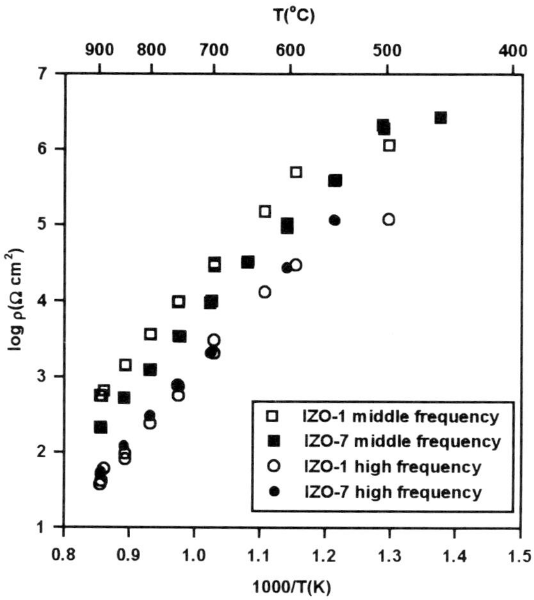

Figure 11 Electrode resistivities of IZO-1 and IZO-7.

8, 9 and 10. This satellite has also very small change for the composition of IZO. $InO_{1.5}$-7 mol% ZrO_2 (IZO-7) was also made by the same route as that of IZO-1. Figure 11 shows there is change for the resistivity of the middle frequency range but not of the high frequency region. The satellite of the high frequency range may be mainly due to charge transfer reaction on the interface of IZO-1 and 8YSZ involving the oxygen vacancy $V_o^{\cdot\cdot}$ in 8YSZ. The capacity of the satellite is about 10^{-6} F/cm^2 independent of P_{O_2} as shown in Fig. 10. The number of charge carriers, i.e. oxygen vacancies, can be obtained as about 3×10^{10} ions/cm^2 by calculation with the AC voltage amplitude of 10 mV and the capacity of the high frequency satellite. Theoretical oxygen vacancy concentration of 8 YSZ is about 2×10^{21}/cm^3 assuming no defect cluster. If these vacancies are uniformly distributed in the 8YSZ electrolyte, about 2×10^{14} ions/cm^2 exist on the interface of 8YSZ and IZO-1, 10 000 times larger than the value calculated from the capacity. The characteristic relaxation time constant is about 2×10^{-5} s. The resistivity of the satellite at the middle frequency range has $P_{O_2}^{-0.5}$ dependence, showing a process involving atomic oxygen. This may be attributed to the oxygen diffusion to the interface of IZO-1 and 8YSZ. The results agree with those of Pt electrode on stabilised zirconia reported by Verkerk *et al.*[10] The capacity of the satellite is about 10^{-4} F/cm^2, almost independent of P_{O_2}, as shown in Fig. 10. Verkerk *et al.* mentioned the relationship between the $\omega_m(=1/\tau)$ and D_O (the diffusion coefficient of oxygen atom) as:

$$\frac{D_O}{d^2} = \frac{\pi\omega_m}{8} = \frac{\pi}{8\tau} \tag{4}$$

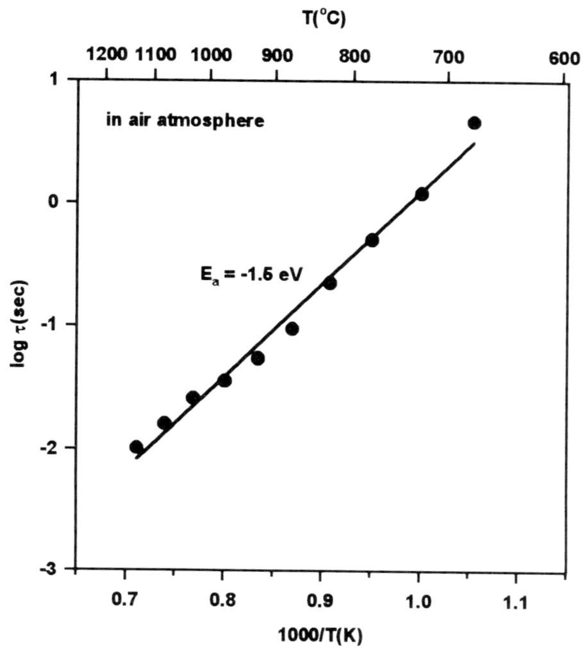

Figure 12 Temperature dependence of relaxation time constant of middle frequency.

where d is the diffusion distance of oxygen at phase three boundary of gas, electrode, and electrolyte.

The characteristic relaxation time constant, $\tau(\sim(D_O/d^2)^{-1})$, varies with $\sim P_{O_2}^{-0.5}$. Wirtz *et al.*[6] measured the diffusion coefficients of oxygen in In_2O_3 thin film vs. partial oxygen pressure by using YSZ as a blocking electrode to electron. Their result shows at high P_{O_2} region ($P_{O_2} > 10^{-4}$ atm) the diffusion coefficient has $P_{O_2}^{+0.5}$ relation by the oxygen interstitial mechanism. From this, diffusion distance, d, seems to be hardly dependent on P_{O_2}. Diffusion distance calculated from the Arrhenius graph of the diffusion coefficient of oxygen atom in IZO-1 (Fig. 2) and τ (Fig. 11) is about 0.2μ, almost independent of temperature as seen in Fig. 13.

Complex impedances were also measured at various temperatures to get activation energies with a new symmetric sample. The activation energy of a resistivity at each frequency range had similar values as seen in Fig. 14.

4. CONCLUSIONS

1. The thermal expansion coefficients of IZO-1 and 8YSZ are 8.826×10^{-6} and 9.721×10^{-6}, respectively. IZO-1 is chemically inert with 8YSZ in the accuracy of XRD.
2. Diffusion coefficient and surface exchange coefficient of oxygen at IZO-1 are $D = 0.1 \exp(-1.8\,\text{eV/kT})$ [cm^2/s] and $k = 1.3 \times 10^{-5} \exp(-0.4\,\text{eV/kT})$ [cm/s].

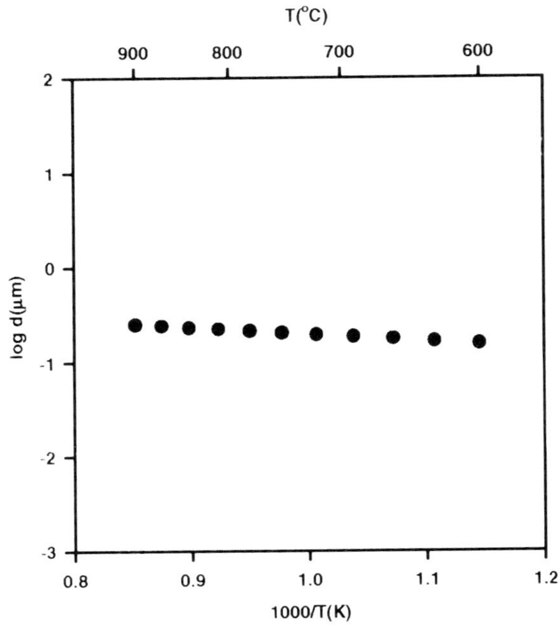

Figure 13 Temperature dependence of diffusion distance.

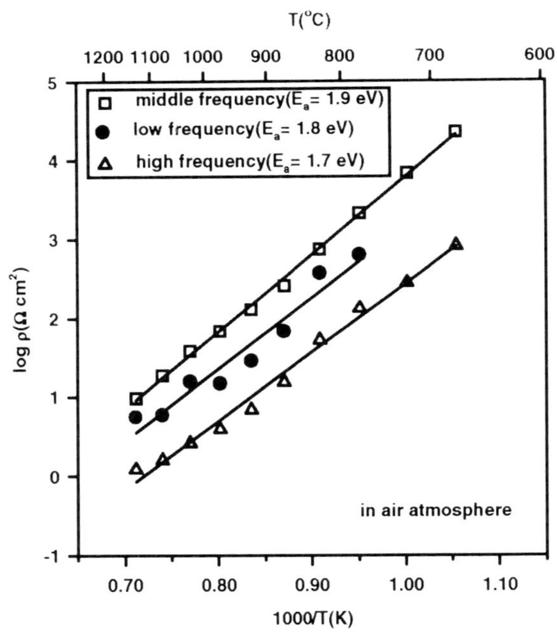

Figure 14 Temperature dependence of resistivity.

3. Impedance response of IZO-1 electrode on 8YSZ mainly consists of three semicircles. At the low frequency region, the resistivity, characteristic relaxation time constant and capacity have $\sim P_{O_2}^{-1}$, $P_{O_2}^{0}$, and $P_{O_2}^{+1}$, respectively. The activation energy of resistivity at the low frequency range is 1.8 eV. The impedance response at this range may be related to gas phase diffusion and adsorption of oxygen from P_{O_2} dependence.
4. At the middle frequency range, the resistivity, characteristic relaxation time constant and capacity have $\sim P_{O_2}^{-0.5}$, $P_{O_2}^{-0.5}$, and $P_{O_2}^{0}$, respectively. The activation energies of the resistivity and relaxation time constant at the middle frequency region are 1.9 eV and 1.5 eV. The electrodic reaction in this range may be related to oxygen diffusion in the electrode.
5. Impedance responses at the high frequency region were independent of oxygen partial pressure and composition of IZO, although resistivity of the middle frequency region was decreased by composition change from IZO-1 to IZO-7. The activation energy of resistivity of this frequency range is 1.7 eV. High frequency response may be due to charge transfer reaction at the interface of IZO-1 and 8YSZ.

ACKNOWLEDGEMENT

This work was financially supported by Korean government.

REFERENCES

1. B. C. H. Steele: *Mat. Sci. Engng*, 1992, **B13**, 79–87.
2. D. Braunshtein, D. S. Tannhauser and I. Riess: *J. Electrochem. Soc.*, 1981, **128**(1), 82–89.
3. Y. Takeda, R. Kanno, M. Noda, Y. Tomida and O. Yamamoto: 1987, **134**(11), 2656–2661.
4. K. Sasaki, H. P. Seifert and L. J. Gauckler: *J. Electrochem. Soc.*, 1994, **141**(10), 2759–2768.
5. J. Crank: *The Mathematics of Diffusion*, 2nd edn, Oxford University Press, 1975.
6. G. P. Wirtz and H. P. Takiar: *J. Am. Ceram. Soc.*, 1981, **64**(12), 748–752.
7. J. Berger, I. Riess and D. S. Tannhauser: *Solid State Ionics*, 1985, **15**, 225–231.
8. D. W. Strickler and W. G. Carlson: *J. Am. Ceram. Soc.*, 1964, **47**(3), 122–127.
9. N. Wakabayashi and O. Yamamoto: *Solid State Ionics*, 1988, **28–30**, 1510–1513.
10. J. Verkerk and A. J. Burggraaf: *J. Electrochem. Soc.*, 1983, **130**(1), 78–84.

Survey of Materials Selection for Ceramic Fuel Cells: I. Electrolytes and Bi-polar Plates

BRIAN C. H. STEELE
Department of Materials, Imperial College, London SW7 2BP, UK

ABSTRACT
By assigning target values ($0.15\Omega\ cm^2$) for the area specific resistance of the individual components in a single SOFC cell it is possible to identify three operating temperature regimes: 500–700°C; 700–900°C; 900–1000°C; for SOFC stacks. These are determined by the relevant properties of the bi-polar plate (interconnect) material, and the appropriate ionic conductivity values for the ceramic electrolyte. Particular attention is given to the level of electronic conductivity in CeO_2 based electrolytes, and to the width of the ionic domain in $La_{0.9}Sr_{0.1}Ga_{0.8}Mg_{0.2}O_{2.85}$ materials.It is emphasised that SrO and CaO doped $LaCrO_3$ bi-polar plate materials are not expected to be stable in the CO_2 partial pressures prevailing at 85% fuel conversion levels. Moreover these compositions exhibit dimensional instabilities in the oxygen partial pressure gradients experienced in SOFC stacks. Problems have also been encountered with metallic alloy bi-polar plates due to the high vapour pressures of chromium containing gaseous species. It follows that there is increasing interest in operating SOFC stacks at intermediate temperatures (500–700°C) using ferritic stainless steel bi-polar plates.

1. SOFC PERFORMANCE AND MATERIALS SELECTION

The design process, which is usually initiated by a perceived market opportunity, is a complicated iterative procedure from the initial conceptual stage to final product.[1]

At the conceptual stage it is desirable to adopt appropriate performance and cost criteria.Typical performance targets for single cells are about $0.5\ Wcm^{-2}$ at 0.7 V and $0.7\ Acm^{-2}$ for 85% fuel conversion. It is recognised that these performance values will be lower in stacks due to additional resistive losses. Neither is it realistic to propose very high power density specifications as these are unlikely to be achieved in practice, and in any case would introduce additional complications for thermal management. The modified Ellingham diagram, Fig. 1, incorporates a line a – – – – a which indicates how the equilibrium oxygen partial pressure varies with temperature for 85% utilisation at 1 bar for an H_2/CO fuel having an H/C ratio of 4:1. It is immediately apparent that at high temperatures (~1000°C) the total polarisation* ($IR + \eta_c + \eta_a$) cannot exceed about 0.15 V if the target output voltage of 0.7 V is to be attained. At 700°C the Nernst open-circuit voltage is about 0.95 V

*At current densities of $0.7\ Acm^{-2}$ concentration polarisation is usually not important with well-designed systems and electrodes, and so it it is only necessary to consider electrolyte losses (IR), together with cathodic (η_c) and anodic (η_a) overpotentials.

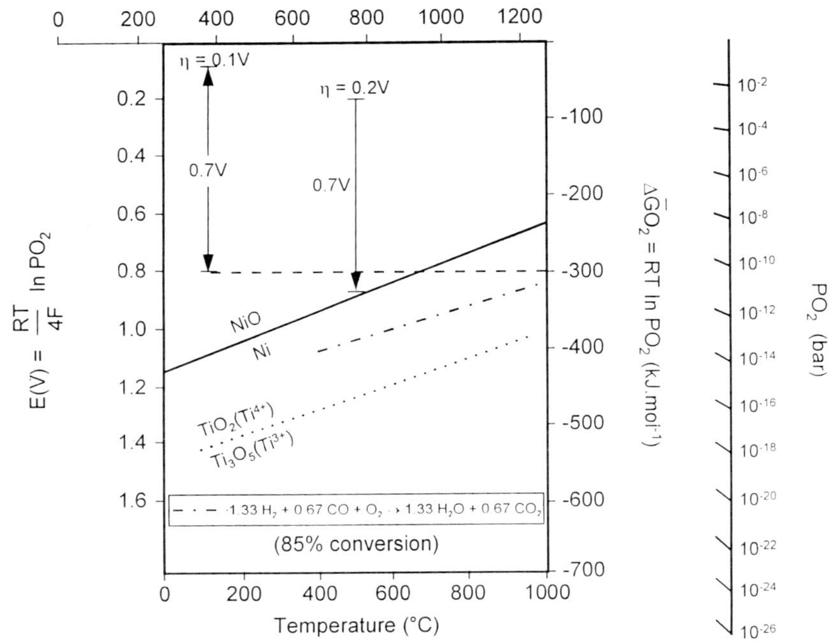

Figure 1 Modified Ellingham diagram depicting typical SOFC operating parameters for 85% fuel conversion.

thus allowing approximately 0.25 V polarisation losses, and obviously higher losses can be tolerated at lower temperatures. For the preliminary evaluation in the present survey a total individual cell polarisation of 0.315 V has been assumed. For current densities of 0.7 Acm^{-2} this implies an area specific resistivity (ASR) for the cell of $0.45\Omega\, cm^2$. This ASR value has been arbitrary divided equally between the cell components, ie electrolyte ($0.15\Omega\, cm^2$), cathode ($0.15\Omega\, cm^2$), and anode ($0.15\Omega\, cm^2$). These values will be used for the various component selection exercises in the subsequent sections, but obviously other resistivity criteria could be used as appropriate.

Another paper[2] considers materials selection criteria for cathodes and anodes, and the present contribution is concerned solely with ceramic electrolytes and bi-polar plate materials.

2. SOLID OXIDE FUEL CELLS

2.1 General Features

Compared to other fuel cells, the solid oxide fuel cell (SOFC) possesses many advantages in that it is a two phase gas/solid system. This overcomes many of the problems associated with liquid electrolytes such as corrosion, flooding, electrolyte distribution, and maintaining a stable Triple Phase Boundary (TPB) region.

Moreover ceramic ion conducting electrolytes can in principle operate over a wide range of temperatures: 500–1000°C. It is useful to designate three operating temperature regimes which are determined by the relevant properties of the bi-polar plate (interconnect) material, and/or the behaviour of the associated ceramic electrolyte material. These regimes are shown in Fig. 2 in which it has been assumed[3] that the area specific resistance of the electrolyte should be less than $0.15\Omega\,cm^2$ for a planar configuration. This area specific resistance is equal to L/σ where L (cm) is the thickness of an oxide ceramic exhibiting a specific ionic conductivity value of $\sigma\,Scm^{-1}$. Present day thick film ceramic technology can provide reliable material down to approximately 10–15 μm which restricts the use of ZrO_2 based electrolytes to temperatures above approximately 700°C and CeO_2 based electrolytes to temperatures above 450°C. If the lower temperature limits for the electrolytes and $LaCrO_3$ based ceramic bi-polar plate are combined with the probable upper temperature limit for bi-polar plates incorporating metallic alloys (see Fig. 2) then three operating temperature regimes can be designated as follows:

(1) 900–1000°C: all ceramic fuel cells.
(2) 700–900°C : supported ZrO_2 electrolytes in association with special metallic (e.g. Siemens) or cermet (e.g. Tonen) bi-polar plate materials
(3) 450–700°C : CeO_2 based electrolytes in conjunction with ferritic stainless steel bi-polar plate materials.

Attention is drawn to the electrolyte, acceptor doped $LaGaO_3$ which is included in Fig. 2.

Although this material has a high ionic conductivity,[4,5] there remain uncertainties about the width of its ionic domain due to the possible formation of Ga^+ under reducing conditions. If these doubts can be eliminated then there may be a role for this electrolyte in SOFC technology in spite of the high cost of Ga_2O_3.

2.2 Operation in 900–1000°C Range

A useful survey of conventional high temperature SOFC technology has been provided by Minh.[6] For SOFC operation at 950–1000°C the following materials have been optimised over the past 25 years and shown to be compatible with little degradation after more than 40,000 hrs (5 yrs) use, providing that processing conditions and compositions are selected to avoid the formation of the interfacial reaction products, $La_2Zr_2O_7$, and $SrZrO_3$.

Electrolyte	: $Zr_{0.835}Y_{0.165}O_{1.917}$ (ZrO_2/Y_2O_3, 91/9)
Cathode	: $La_{0.8}Sr_{0.2}MnO_{3+x}$
Anode	: Ni–$ZrO_2(Y_2O_3)$
Interconnect	: $LaCr_{0.8}Mg_{0.2}O_3$

Although there is a general consensus regarding the selection of these materials

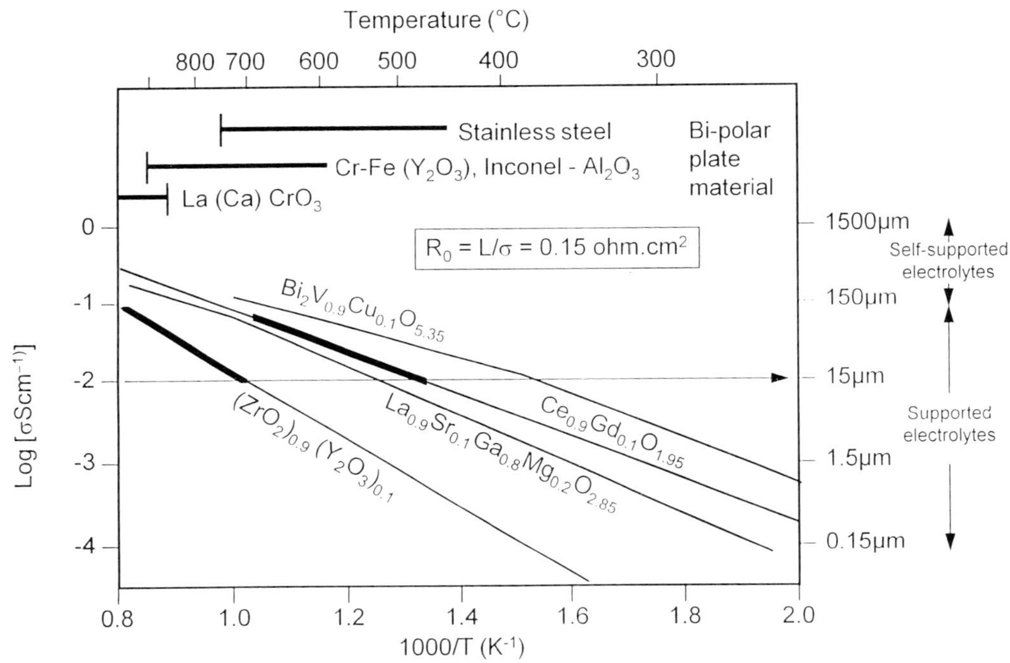

Figure 2 Specific ionic conductivity values for selected oxide electrolytes as a function of reciprocal temperature.

there is still a divergence of opinion about the best configuration to be adopted that will allow the economical mass production of reliable, strong tough ceramic components. Westinghouse have optimised a tubular arrangement which has been scaled up to 25 kW prototype units now under evaluation by gas and electric utilities. A porous cathode ($La_{0.8}Sr_{0.2}MnO_3$) support tube approximately 1 cm diameter and 1 m in length is first fabricated by conventional ceramic processing routes. Impermeable interconnect (40 μm) and electrolyte (40 μm) components are deposited sequentially by either ECVD or plasma-spraying techniques onto the porous cathode support tube. Finally a porous Ni–ZrO_2 anode coating is produced by a combination of slurry deposition and CVD. The individual tubular cells are assembled into a bundle with nickel felt current collectors providing appropriate series-parallel connections. The Westinghouse design avoids the necessity of having high temperature seals, and, for in-situ reforming operation, the fuel/steam supply can be distributed along the length of the tubular cell to reduce thermal stresses arising from the endothermic reforming reaction. Moreover the tubular module incorporates a ceramic heat exchanger so that the temperature of the exhaust gas is approximately 850°C which allows the use of conventional stainless steel ducting, pumps, etc. for the 'balance-of-plant' equipment. The size of the unit is being increased and Westinghouse expect to produce a 100 kW prototype for evaluation in the Netherlands by 1997. The principal queries associated with the Westinghouse configuration are concerned with the scale-up of the fabrication techniques employed and the low power density of the design. The relatively slow ECVD process, and associated high temperature masking procedures, introduce a major cost penalty. Selected groups in Japan are trying to reduce the processing costs by using more conventional ceramic fabrication routes, but it is proving very difficult to co-sinter the doped $LaCrO_3$ interconnect and zirconia electrolyte components.

2.3 Operation in 700–900°C Range

In principle, the planar configuration, which is usually adopted for liquid electrolyte fuel cell systems, offers many advantages over the tubular design. A planar arrangement should develop higher power densities, and fabrication of the cell components can be accomplished by mass-production methods such as tape casting, calendering, or screen-printing. However, the practical realisation of the planar configuration requires the individual cells to be sealed at the edges. Moreover, it is doubtful whether the electrolyte component (150–200 μm thick) can be reliably and cheaply fabricated at sizes much greater than 15 × 15 cm. The economical manufacture of doped $LaCrO_3$ bi-polar (interconnect) plates also presents a problem as this material is difficult to sinter except at higher temperatures (~1700°C) in reducing atmospheres, and liquid phase sintering can introduce long term stability problems. It appears that only the MgO doped composition is stable, as the SrO and CaO doped material can form carbonates, and also exhibit significant stresses due to lattice expansions associated with the loss of oxygen under anodic conditions (see Section 6.1). Fortunately, the recent development by Siemens (e.g. ref. 7) of a metallic bi-polar plate has overcome many of the difficulties. A metallic 'window-frame' design allows both the use of relatively small area (10 × 10 cm) electrolyte–electrode PEN* structures, which are in contact with large area metallic bi-polar plates. The introduction of a metallic framework and bi-polar plate introduces many benefits. For example, the superior thermal conduction of the metallic component reduces thermal gradients and improves the efficiency of the cooling fluid which is usually excess air in the cathode channels. This concept has so far (July 1995) been scaled up to 10 kW units.

2.4 Operation in 500–700°C Range

At these temperatures it becomes possible to specify ferritic stainless steel compositions for the bi-polar plate. These alloys have appropriate thermal expansion coefficients and oxidation resistance, although at present there is little information about interfacial electrode/bi-polar plate contact resistances. At 700°C excellent performances (~0.5 Wcm^{-2}) have been reported[8] for the cell:

$$La_{0.6}Sr_{0.4}Co_{0.2}Fe_{0.8}O_{3-x}/Ce_{0.9}Gd_{0.1}O_{1.95}/Ni\text{-}ZrO_2(CeO_2).$$

*PEN: Positive-electrolyte-negative.

Projects are underway to optimise these structures and to evaluate the performance of small stacks. An outstanding question concerns the development of electronic conductivity in ceria based electrolytes (see Section 4.1), and whether it will be necessary to coat the anode side with an electronic blocking material. Attempts are also being made to fabricate electrode supported $Ce_{0.9}Gd_{0.1}O_{1.95}$ electrolytes (10 μm thick) which can be incorporated into stacks designed to operate at 500°C in electric vehicles supplied with methanol or reformate fuel.

3. ZrO_2 SOLID SOLUTIONS

Although operation at elevated temperatures (900–1000°C) is well documented[6] there is little information about the use of zirconia electrolytes at intermediate temperatures. Examination of Fig. 2 indicates that to meet the target resistance value of 0.15Ω cm^2 at 700°C zirconia electrolytes have to be fabricated as 15 μm thick films. Similar thicknesses are required for both cubic and tetragonal electrolytes. Although *t*–ZrO_2 electrolytes are stronger at room temperature there appears to be no reason to select this material as the supporting electrode is the principal structural component, and in any case the transformation toughening mechanism is not operative at elevated temperatures. Moreover, there is some evidence[9] that *t*–ZrO_2 electrolytes can de-mix after the prolonged passage of high currents resulting in the formation of high resistance monoclinic ZrO_2 material. Also additions of Al_2O_3 which are sometimes added to improve the toughness of ZrO_2 electrolytes[10] appear an unnecessary complication for thick film electrolytes. Unless the oxygen reduction kinetics at intermediate temperatures can be improved for ZrO_2 electrolyte/electrode combinations there does not appear to be much incentive to use simple electrode supported ZrO_2 electrolytes as the performance of CeO_2 electrolyte/electrode combinations are superior. The use of thick film ZrO_2 electrolytes on CeO_2 substrates also appears problematical in view of the thermal expansion mismatch.

In general ZrO_2/Y_2O_3 ratios of 91/9 mol% are favoured for operation in the range 900–1000°C to ensure that the decrease in ionic conductivity due to vacancy ordering and formation of second phase is minimised. For operation at 700°C there appears to be no long term operational evidence that favours either the 8 or 9 mol% Y_2O_3 dopant values as both ‘cubic’ compositions are metastable at this temperature.

It is known that ZrO_2–Sc_2O_3 solid solutions[11] exhibit greater ionic conductivity values than ZrO_2–Y_2O_3 compositions. In thick film form the high cost of Sc_2O_3 would not be such a disadvantage and there may be designs/applications where further consideration of this material might be appropriate.

Finally, the concept of fabricating single phase p-i-n zirconia based electrolyte/electrode structures has been proposed (eg. ref. 12) using Tb_2O_3 and TiO_2 additions. However these designs are unlikely to perform well without the additions of further porous electrode layers to improve the relatively low oxygen surface exchange values[13] of the doped zirconia surfaces.

4. CeO_2 SOLID SOLUTIONS

4.1 General Comments

The superior ionic conductivity exhibited by CeO_2 based electrolytes was known over thirty years ago. Moreover Takahashi[14] demonstrated the excellent electrode kinetics associated with CeO_2 electrolyte/electrode combinations, and these observations were confirmed by the oxygen self diffusion (D^*) and surface exchange (k) values measured by Floyd and Steele.[15] Because pure CeO_2 already exhibits the fluorite structure it is not necessary to add a dopant to stabilise the cubic structure, and so the composition can be optimised for maximum ionic conductivity which is usually attained for additions of around 10 mol% of $MO_{1.5}$ to yield oxygen vacancy concentrations of 2.5% (ie $Ce_{0.9}M_{0.1}O_{1.95}$). In contrast the favoured zirconia composition is ZrO_2/Y_2O_3, 91/9, mol%, ie $Zr_{0.835}Y_{0.165}O_{1.917}$ corresponds to an oxygen vacancy concentration of 4.1%, which is already in the region where vacancy ordering will occur.

The optimal dopant is Gd^{3+} ($\Delta E \approx 0.7$ eV) closely followed by Sm^{3+} ($\Delta E \approx 0.72$ eV) to ensure that V_O''–M_{Ce}' interactions are minimised.[16,17] Although there are claims in the literature suggesting alternative dopants these conclusions are based on the interpretations of total conductivity measurements which include a grain boundary resistance. This parameter can be relatively large due to segregation of impurities and dopants but can be reduced to acceptable values.[8] It is interesting to note that two recent papers on grain boundary impedances in $Ce(Gd)O_{2-x}$[18] and $Zr(Y)O_{2-x}$[19] confirm that even with 'pure' materials there appears to be an intrinsic grain boundary impedance.

The principal uncertainty arising from the selection of $Ce_{0.9}Gd_{0.1}O_{1.95}$ (or $Ce_{0.9}Sm_{0.1}O_{1.95}$) electrolytes for use at intermediate temperatures (500–700°C) arises from the loss of oxygen at the oxygen partial pressures prevailing under anodic conditions according to the reaction,

$$O_O^{\times} \rightarrow 1/2O_2 + V_O^{\cdot\cdot} + 2e' \quad \ldots\ldots\ldots\ldots\ldots \mathbf{i}$$

The increase in n-type conductivity which depends upon $P_{O_2}^{-1/4}$, reduces the ionic transference number and associated cell efficiency (eg refs 20, 21, 22). The formation of additional anion vacancies also produces a lattice expansion which may have a deleterious effect on the relevant mechanical behaviour (see Section 6.1 on doped $LaCrO_3$). These two problems are discussed in more detail in subsequent sections.

4.2 Electronic Transference Numbers

Providing t_{ion} exceeds 90% under the operational anode environment then Steele[3] has argued that the open circuit voltage and cell efficiency will only be reduced by a small amount which is generally acceptable particularly as the anode kinetics are usually enhanced under these conditions. Application of this criterion will obviously

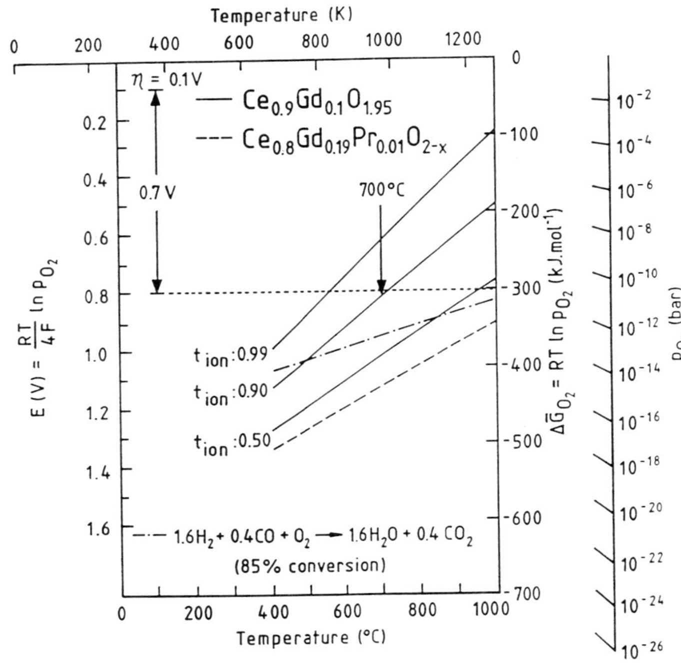

Figure 3 Modified Ellingham diagram incorporating ionic transference number (t_{ion}) data for $Ce_{0.9}Gd_{0.1}O_{1.95}$ and $Ce_{0.8}Gd_{0.19}Pr_{0.01}O_{2-x}$.

depend upon details of the cell operation, and the availability of reliable values of t_{ion}. To calculate the latter it is important to know at the required temperature the value of P_θ, which is the oxygen partial pressure at which $\sigma_{ion} = \sigma_e$. Data assembled by Tuller[23] and recent measurements at Imperial College[8] enabled t_{ion} values to be calculated and discussed,[3] which suggested that $Ce_{1.9}Gd_{0.1}O_{1.95}$ could be used effectively at 500°C, but that its use at 700°C was marginal (see also Fig. 3). In general these predictions have been confirmed recently by ECN[24] and Gödickemeier,[25] but some details require further clarification.

Examination of literature values of P_θ indicate a considerable scatter of values for nominally the same composition (eg. ref. 26). Most investigators measure the total conductivity as a function of P_{O_2}, and thus derive values of P_θ, although other techniques are available.[27,28] The total conductivity measurements could include a grain boundary contribution which may complicate the interpretation, and collaborative studies are needed to establish definitive values of P_θ. The electronic conductivity values incorporate the product of two parameters, concentration of charge-carriers (n), and a mobility term (μ),

ie $$\sigma_e = ne\mu.$$

It is to be expected that the value of n, for small dopant levels, will be closely related to the thermodynamics[29] of non-stoichiometric CeO_{2-x}, and so variations of

σ_e probably reflect changes in μ. These variations in μ could arise due to trapping centres in the relevant CeO_{2-x} solid solutions, or due to Schottky barriers at the grain boundaries. Maricle *et al.*[30] claimed that small additions of Pr^{3+} act as trapping centres thus decreasing the level of electronic conductivity and changing the value of P_O. However, Steele *et al.*[8] were unable to confirm this observation although in principle the concept might work with other redox cations.

The selection of CeO_2 based electrolytes for SOFC systems has recently been stimulated by the design concepts outlined by Barp[31] and Bos.[32] These concepts envisage using small (1–2 kW size) SOFC systems as the base load for domestic applications with the peak power demand being supplied by the grid or batteries. These systems would operate at optimal efficiency (>85% fuel conversion) which would minimise any deleterious effects arising from the presence of electronic conductivity.

4.3 *Lattice Expansion*

At the low oxygen partial pressures corresponding to anodic conditions additional anion vacancies are generated according to equation (i). The associated lattice expansion produced by coulombic relaxation around these vacancies will produce strains in an SOFC stack, and calculations for doped $LaCrO_3$ (Section 6.1) suggest that the associated stresses can be significant and might be responsible for mechanical failure. Similar calculations are in progress for $Ce_{0.9}Gd_{0.1}O_{1.95}$,[33] and other CeO_2 based electrolytes, but the effect is likely to be negligible at 500°C.

4.4 *Duplex Electrolytes Incorporating CeO_2*

As the anodic t_{ion} for $Ce_{0.9}Gd_{0.1}O_{1.95}$ will be smaller than the target value of 0.9 for many operational conditions at 700°C it may be necessary to incorporate an electronic blocking layer on the CeO_2 electrolyte adjacent to the anode. This type of structure has been examined both with modelling calculations[22,34] and by experimental measurements.[35,36] Whilst the electronic conductivity is effectively blocked by thin layers of $Zr(Y)O_{2-x}$, the selection of $Zr(Y)O_{2-x}$ is not satisfactory due to the thermal expansion mismatch with $Ce_{0.9}Gd_{0.1}O_{1.95}$. However alternative solid electrolytes are probably available which could exhibit better thermal expansion compatibility with CeO_2 solid solutions, and appropriate measurements are urgently required.

5. ALTERNATIVE SOLID ELECTROLYTES

Many investigations have been concerned with the evaluation of alternative oxide electrolytes which might exhibit superior performance to the well-known ZrO_2 and CeO_2 based materials. Both fluorite related, eg pyrochlore ($A_2B_2O_7$), and perovskite related, eg brownmillerite ($A_2B_2O_5$), have been examined together with doped rare-earth oxides. These have often been reviewed (eg refs 37, 38, 39, 40) and it will be noted that doped La_2O_3 and $Zr_2Gd_2O_7$ have values close to $(ZrO_2)_{0.9}(Y_2O_3)_{0.1}$ at intermediate temperatures, although the possibility of significant protonic conductivity in doped La_2O_3 should not be overlooked.[41]

6. BI-POLAR PLATES

6.1 Doped $LaCrO_3$

During the 1960s and 1970s a number of materials were evaluated by Westinghouse, GEC, and Brown-Boveri[6] as possible bi-polar/interconnect materials with most groups eventually selecting doped $LaCrO_3$ compositions. Twenty years ago Westinghouse decided to use MgO doped $LaCrO_3$ interconnect for their tubular configuration fabricated by the ECVD process. Although the higher vapour pressure of $MgCl_2$ conferred a specific advantage for the MgO dopant in the ECVD process it was not apparent at the time that this advantage was sufficient to overcome the disadvantage of MgO doped $LaCrO_3$ including thermal expansion mismatch, compared to the SrO and CaO doped material. These disadvantages also include relatively poor electronic conductivity and poor sintering behaviour. In the last two years, however, it has been recognised that SrO and CaO doped $LaCrO_3$ are thermodynamically unstable[42] when exposed to the CO_2 partial pressure (>0.1 bar) generated in the anode compartment assuming about 85% fuel conversion. Moreover the relatively large lattice expansions associated with the loss of oxygen under anodic conditions can create significant strains,[43,44,45] particularly for SrO and CaO doped compositions, which can have an adverse effect on the structural integrity of these materials. These recent data imply that only MgO doped $LaCrO_3$ materials can be confidently selected for SOFC applications and it is only possible to use these compositions at high temperatures (>850°C?) due to their low electronic conductivity. It follows that they cannot be recommended for incorporation in intermediate temperature SOFC stacks, particularly as the cost for monolithic components is excessive. Similar comments apply to the composite interconnect/electrolyte concept proposed by ICI.[46] The selection criteria for the bi-polar plate material is very demanding and it is unlikely that an alternative ceramic material will be developed although data reported for novel anode materials should always be evaluated to examine whether there are also possible applications for their use as bi-polar plate components.

6.2 Metallic Bi-polar Plates

The incorporation of a special alloy, Cr–Fe–Y_2O_3 (1 mol%), in the Siemens SOFC configuration[47,7] stimulated investigations into the use of metallic alloys as bi-polar plate material, and Tonen[48] claim success with a cermet having the composition Inconel 600/Al_2O_3 (40/60 vol%). However, both Siemens and Tonen report problems associated with Cr_2O_3 vaporisation which is due to gaseous species such as $CrO_2(OH)_2$ and CrO_3.[42] At lower temperatures (<700°C) it becomes possible to specify ferritic stainless steel[49] which has low creep rates at this temperature, and thermal expansion coefficient values (12–13 ppm) close to CeO_2 based electrolytes. The principal oxidation product is usually $FeCr_2O_4$, and investigations are urgently required to provide data on interfacial contact resistances with the adjoining electrode components over an extended period of time.

Although more expensive, a range of Ti alloys (e.g. Timet, Titanium Metal Corp) has been developed for the aerospace industry for applications up to 600°C. With thermal expansion coefficient values in the range 8.5–10 ppm, appropriate designs could ensure that ceramic components are maintained under compressive stresses. Any evaluation of titanium alloys for intermediate temperature SOFC application will also have to pay particular attention to effects arising from the possible dissolution of hydrogen from the anode compartment.

7. ACKNOWLEDGEMENTS

The author wishes to thank present colleagues in the Centre for Technical Ceramics for many useful discussions and for providing experimental data. Financial support from EPSRC, Commisson of the European Communities, and a NEDO International Joint Research Grant, is also gratefully acknowledged.

8. REFERENCES

1. M. F. Ashby: *Mat. Sci. and Tech.*, 1989, **5**, 517.
2. B. C. H. Steele: 'Proc. of Solid State Ionics – 10', Singapore, 3–8 Dec., 1995. To be published in *Solid State Ionics* 1996.
3. B. C. H. Steele: p. 375, *Proc. of 1st European SOFC Forum*, U. Bossel (ed.), 1994.
4. T. Ishihara, H. Minami, H. Matsuda and Y. Takita: *Solid Oxide Fuel Cells IV*, M. Dokiya *et al.* (eds), *Proc. Electrochem Soc.*, 1995, **95**(1), 344, New Jersey, USA.
5. M. Feng and J. B. Goodenough: *Eu. J. Solid State Inorg. Chem.*, 1994, **31**, 663.
6. N. Q. Minh: *J. Am. Ceram. Soc.*, 1993, **76**, 563.
7. L. Blum, W. Drenckhahn, H. Greiner and E. Ivers-Tiffe: *Solid Oxide Fuel Cells IV*, M. Dokiya *et al.* (ed.), *Proc. Electrochem Soc.*, 1995, **95**(1), 163, New Jersey, USA.
8. B. C. H. Steele, K. Zheng, R. A. Rudkin, N. Kiratzis and M. Christie: *Solid Oxide Fuel Cells IV*, M. Dokiya *et al.* (eds), *Proc. Electrochem Soc.*, 1995, **95**(1), 1028. New Jersey, USA.
9. S. P. S. Badwal and N. Nardella: *Applied Physics A*, 1989, **A49**, 13.
10. W. Schafer, H. Geier, G. Lindemann and D. Stolten: 'High Temperature Electrochemical Behaviour of Fast Ion and Mixed Conductors', F. W. Poulsen *et al.* (eds), *Proc. 14th RISO Symp.*, 1993, 409, RISO Nat. Lab., Roskilde, Denmark.
11. Y. Mizutani, M. Tamura, M. Kawai, K. Nomura, Y. Nakamura and O. Yamamoto: *Solid Oxide Fuel Cells IV*, M. Dokiya *et al.* (eds), *Proc. Electrochem. Soc.*, 1995, **95**(1), 30, New Jersey, USA.
12. W. L. Worrell, P. Han, Y. Uchimoto and P. K. Davies: *ibid.*, p. 50.
13. B. C. H. Steele: *Solid State Ionics*, 1995, **75**, 157.
14. T. Takahashi: *Physics of Electrolytes*. 1972, **2**, J. Hladik (ed.), Academic Press.
15. B. C. H. Steele and J. M. Floyd: *Proc. Brit. Ceram. Soc.*, 1971, **19**, 55.
16. *Non-Stoichiometric Oxides*, O. Toft Sorensen (ed.), 1981, Academic Press.

17. A. S. Nowick: *Diffusion in Crystalline Solids*, G. E. Murch and A. S. Nowick (eds), 1984, 143, Academic Press.
18. G. M. Christie and F. P. F. van Berkel: Accepted for publication in *Solid State Ionics*, 1996.
19. A. Aoki, Y.-M. Chiang, I. Kosacki, J.-R. Lee, H. L. Tuller and Y. Liu: Accepted for publication in *J. Am. Ceram. Soc.*, 1996.
20. H. Schmalzried: *Z. Phys. Chem. N.F.*, 1963, **38**, 87.
21. B. N. Ross and T. G. Benjamin: *J. Power Sources*, 1976/77, **1**, 311.
22. I. Riess: *Solid State Ionics*, 1992, **52**, 127.
23. H. L. Tuller: *Non-Stoichiometric Oxides*, O. Toft Sorensen (ed.), 1981, 271, Academic Press.
24. F. P. F. van Berkel, G. M. Christie, F. H. van Heuveln and J. P. P. Huijsmans: *Solid Oxide Fuel Cells IV*, M. Dockiya *et al.* (eds), *Proc. Electrochem. Soc.*, 1995, **95**(1), 1062, New Jersey, USA.
25. M. Gödickemeier, K. Sasaki and L. J. Gauckler: *ibid.*, 1072.
26. M. Mogensen, T. Lindegaard, U. R. Hansen and G. Mogensen: *Ionic and Mixed Conducting Ceramics II*, T. A. Ramanarayanan *et al.* (eds), *Proc. Electrochem. Soc.*, 1994, **94**(12), 448, New Jersey, USA.
27. W. Weppner and R. A. Huggins: *Ann. Rev. Mater. Sci.*, 1978, **8**, 269.
28. F. Marques: University of Aveiro, Private communication, July, 1995.
29. B. Zachau-Christiansen, T. Jacobsen, K. West and S. Skaarup: *Solid Oxide Fuel Cells III*, S. C. Singhal and H. Iwahara (eds), *Proc. Electrochem. Soc.*, 1993, **93**(4), 104, New Jersey, USA.
30. D. L. Maricle, T. E. Swarr and S. Karavolis: *Solid State Ionics*, 1992, **52**, 173.
31. P. Bos: *Proc. 4th Grove Fuel Cell Symposium*, 19–22 Sept., 1995, to be published in *J. Power Sources*, 1996.
32. B. Barp, R. Diethelm, K. Honegger and E. Batawi: *ibid.*
33. A. Atkinson: Imperial College, Private communication, Aug. 1995.
34. A. V. Virkar: *J. Electrochem. Soc.*, 1991, **138**, 1481.
35. K. Mehta, S. J. Hong, J.-F. Jue and A. V. Virkar: *Solid Oxide Fuel Cells III*, S. C. Singhal and H. Iwahara (eds), *Proc. Electrochem. Soc.*, 1993, **93**(4), 92, New Jersey, USA.
36. H. Yahiro, Y. Baba, K. Eguchi and H. Arai: *J. Electrochem. Soc.*, 1988, **135**, 2077.
37. B. C. H. Steele: *High Conductivity Solid Ionic Conductors*, 1989, 402, T. Takahashi (ed.), World Scientific Publ., Singapore.
38. B. C. H. Steele: *Mat. Sci. & Eng.*, 1992, **B13**, 79.
39. A. J. Burgraff, B. A. Boukamp, I. C. Vinke and K. J. de Vries: *Advances in Solid State Chemistry*, 1989, **1**, 259.
40. R. I. Cook and A. F. Sammells: *Solid State Ionics*, 1991, **45**, 311.
41. Y. Larring and T. Norby: *Solid State Ionics*, 1995, **77**, 147.
42. D. H. Peck, M. Miller, H. Nickel, D. Das and K. Hilpert: *Solid Oxide Fuel Cells IV*, M. Dockiya *et al.* (eds), *Proc. Electrochem. Soc.*, 1995, **95**(1), 858, New Jersey, USA.
43. P. V. Hendrickson, J. D. Carter and M. Morgensen: *ibid*, 934.
44. I. Yasuda and M. Hishinuma: *ibid.*, 924.
45. C. S. Montross: Accepted for publication in *Brit Ceram. Trans.*, 1995.
46. N. Alford, J. D. Birchall, A. J. Howard, K. Kendall and J. H. Raistrick: ICI Patent Application EPO1834453.
47. W. Drenckhahn: *Solid Oxide Fuel Cells II*, F. Grosz *et al.* (eds), EUR 13564 EN (Direct.-Gen XII), 43, Brussels, 1991.
48. H. Seto, T. Miyata, A. Tsunoda, T. Yoshida and S. Sakurada: *Solid Oxide Fuel Cells III*, S. C. Singhal and H. Iwahara (eds), *Proc. Electrochem. Soc.*, 1993, **93**(4), 421, New Jersey, USA.
49. W. F. Smith: *Structure and Properties of Engineering Alloys*, 2nd edn, McGraw Hill, 1993.

Improving Electrode Contact in a Small Diameter Tubular SOFC

N. J. COE, K. KENDALL, I. KILBRIDE and M. PRICA
Centre for Inorganic Chemistry and Materials Science, Keele University, Staffordshire ST5 5BG, UK

ABSTRACT
A major goal in solid oxide fuel cell (SOFC) research is to reduce the resistance of both cell components and interfaces. While this has been addressed in planar designs, little attention has been given to the application of electrodes in small diameter tubular SOFCs, which show inherent advantages in terms of thermal shock resistance, sealing and cost. Nickel metal as an anode material gives a high in-plane conductivity but because of poor adhesion peels away from the zirconia electrolyte, giving a high interfacial resistance. Improved adhesion between the anode and electrolyte was achieved by modifying the electrolyte surface with a nitrate salt solution. Current density was improved by a factor of two by applying $Ni(NO_3)_2$ and $Mn(NO_3)_2$ to the inner and outer tube walls respectively. This increase was attributed to improved anode electrochemical performance due to enhanced cohesion between nickel metal and the zirconia surface. There was some evidence for the formation of an intermediate phase at the interface in $Ni(NO_3)_2$-treated tubes.

1. INTRODUCTION

Unlike other designs, small diameter tubular solid oxide fuel cells have excellent thermal shock characteristics and are resistant to high thermal gradients. Measurements can be made in minutes rather than hours and fuel pipes can be sealed outside the furnace hotzone making high temperature sealing redundant.[1]

NiO/yttria-stabilised zirconia (YSZ) cermet is the material used in contemporary SOFCs because it has good catalytic and electrical characteristics and has a compatible thermal expansion coefficient with zirconia at the cell operating temperature. There are difficulties, however, associated with the preparation and the application of NiO/YSZ cermets for the tubular SOFC. It is easy to obtain a fine micron sized nickel powder but NiO/YSZ cermets are more difficult to grind and apply. Also, because of the longer current path associated with the tubular cell, it is advantageous to use a pure nickel anode for increased in-plane conductivity. Additionally, a nickel anode can be fired when first used, removing a fabricating step.

Although pure nickel metal anodes have these advantages, NiO/YSZ cermets are preferred because pure nickel does not wet zirconia very well. Nickel anodes tend to peel away from the zirconia surface during normal operation and during temperature cycling. Much of the electrical losses in SOFCs are attributable to the electrode interface, commonly called the triple phase boundary (TPB).[2] Clearly if a process were to improve the interaction between nickel and zirconia at the interface the resulting cell would have increased performance. More importantly, an additive at the interface may react with the zirconia, allowing nickel to adhere more readily.

Pyrolysis of suitable metal salts on the zirconia surface followed by application of nickel metal is one method which may improve adhesion. Results which demonstrate this theory have been well documented before using flat plates. Thampi *et al.*[3] have reported improvements in cell performance with Ni and Mn nitrate impregnation. This work applies their method to small diameter tubular cells.

2. EXPERIMENTAL

A tape containing 50 g Unitec FYT11 ZrO_2, 2.7 g PVA, 1.7 g methyl cellulose, 6.0 g of 10% glycerol solution, was made by twin roll milling. This tape was rolled and extruded into tubes which were fired at 1500°C for 1 h. The tubes were then cut to an average length of 110 mm using a diamond edged wheel. The fired tubes had an outer diameter of 2.15 mm ± 0.02 mm, an inner diameter of 1.65 mm ± 0.02 mm and a wall thickness of 0.25 mm ± 0.02 mm.

The zirconia tubes were treated with two nitrate salt solutions, one containing 0.9 g BDH Chem. AnalaR $Ni(NO_3)_2.6H_2O$, 0.12 g PVB, 15 g propan-2-ol and the other containing 0.9 g Fluka $Mn(NO_3)_2.6H_2O$, 0.12 g PVB, 15 g propan-2-ol. The $Mn(NO_3)_2$ was used to treat the cathode side whilst $Ni(NO_3)_2$ was used to treat the inner wall (anode side). The tubes were allowed to dry at 100°C and the treatment process was repeated to ensure an adequate and complete coverage. Twelve tubes were prepared in total: 3 tubes were not coated with either nitrate solution; 3 were treated with $Ni(NO_3)_2$ only on the inner wall; 3 tubes were coated with $Mn(NO_3)_2$ only on the outer wall; 3 tubes were coated with both $Ni(NO_3)_2$ on the inner wall and $Mn(NO_3)_2$ on the outer wall. The $Ni(NO_3)_2$-treated tubes were fired at 1420°C for 2 h while the $Mn(NO_3)_2$-treated tubes were fired at 1310°C for 10 min. In the case where both walls were coated, they were fired with $Ni(NO_3)_2$ before $Mn(NO_3)_2$ because of the higher firing temperature required for $Ni(NO_3)_2$.

After nitrate treatment, the inner walls of all the tubes were coated with a nickel metal paste (2.5 g Goodfellows 4–7 μm Ni, 7 g propan-2-ol, 0.65 g PVB). This process was repeated, after drying at 100°C, to ensure a thorough nickel metal coating. A lanthanum strontium manganite (LSM) ink (20 g Merck $La_{0.82}Sr_{0.18}MnO_3$, 0.6 g ICI KD1, 4 g terpineol vehicle) was applied on an area around the circumference of the tubes with an average length of 18 mm giving a surface area of LSM 1.2 cm^2. Platinum wire (0.3 mm diameter) was wrapped around the LSM area for cathode current collection and a nickel wire (0.5 mm diameter) was inserted into the tube for anode current collection (see Fig. 1).

Individual cells were connected to a galvanostat and measured in a purpose-built rig. Hydrogen saturated with 3% water was passed through the cell and the furnace was heated to 1000°C. At this temperature the open circuit voltage (OCV) was recorded and the cell was held at a current of 20 mA for 30 min. After this time the fuel cell was tested at a range of currents to obtain the IV plot. A reference electrode was placed on the cathode side. The voltage difference between these two electrodes

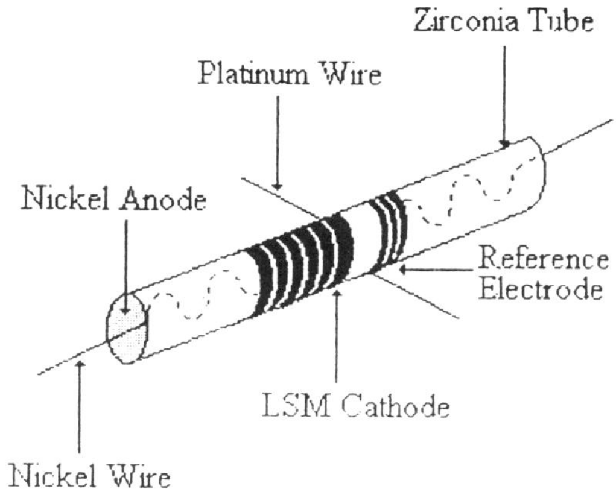

Figure 1 Schematic representation of the fuel cell.

was the cathode 'contribution'. The anode contribution was calculated by subtracting the cathode contribution and the working cell voltage from the OCV. These electrode contributions include the electrode overvoltages, the IR drop due to the resistance of the zirconia and the wire resistances connected to the electrodes.

3. RESULTS AND DISCUSSION

Figure 2 is an IV plot of the average cell performance for the cells treated or untreated with nitrate solutions. The best average cell performance (72 mA cm^{-2} at 0.7 V) was obtained with the combined treatment of $Mn(NO_3)_2$ on the outer walls and $Ni(NO_3)_2$ on the inner walls. A lone coating of $Ni(NO_3)_2$ also appeared to improve current density (37 mA cm^{-2} at 0.7 V) over those measured for the untreated cells (20 mA cm^{-2} at 0.7 V). The increase measured for $Mn(NO_3)_2$-treated cells (25 mA cm^{-2} at 0.7 V) was negligible. The largest current densities obtained with the application of both nitrate salts seem relatively modest when compared with outputs recently achieved.[3] The application of electrodes onto tubes is somewhat more complicated than the application of electrodes onto flat plates. It is difficult to control the uniformity and thickness of the nickel anode applied to the inside of the tube. Additionally, no mechanical pressure can be exerted between the electrodes and wire current collector whilst the cell is in operation. This would be expected to increase contact resistance between the current collectors and the electrodes. It is also expected that the cell output could be increased further by increasing cathode thickness, beyond the 30 μm used in this experiment.

Figure 3(a) shows a plot of the average cell performance for the combined nitrate-treated tubes together with the relative electrode contributions. Figure 3(b) is a similar plot for the untreated tubes with their electrode contributions. In contrast to nitrate-treated tubes, the anodes in untreated tubes provide the most

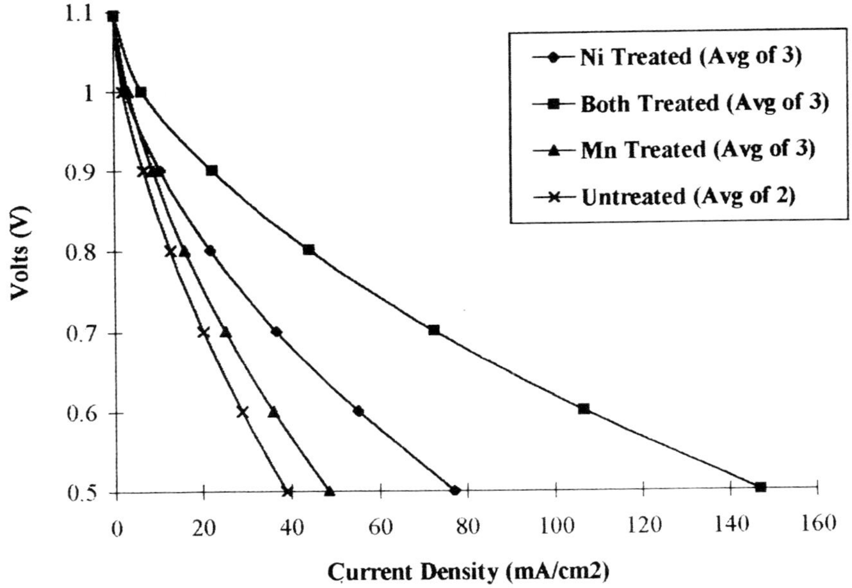

Figure 2 IV plot for cells with various nitrate treatments.

contribution to cell resistance. The cause of this effect is shown in Table 1. The table shows electrode resistances at the working cell voltage of 0.7 V. Nickel nitrate on the inner tube wall lowers the anode contribution to cell resistance whilst simultaneously increasing cell output. No obvious systematic difference is observed between the cathode treatments. It is possible that $Ni(NO_3)_2$ has a synergistic effect upon cell performance in combination with $Mn(NO_3)_2$. Manganese nitrate may improve the cathode interface or stop detrimental reactions between the LSM and zirconia.[4] Alternatively, a catalytic process, such as a redox mechanism, may be responsible for the increase in cell current when both nitrate salts are applied.[3]

Figure 4 shows a SEM micrograph at low magnification of an untreated tube fracture surface cross-section after the cell's operation at 1000°C. The nickel anode has delaminated from the inner tube surface and shows complete interface failure with the zirconia. This behaviour was observed for all untreated tubes. $Ni(NO_3)_2$-treated tubes had continuous adhered nickel anodes after operation at 1000°C which showed no sign of gross delamination. The SEM micrograph also serves to illustrate the difficulty in achieving a coating of even thickness.

Figure 5(a) is a SEM micrograph of a $Ni(NO_3)_2$-treated tube in cross-section showing the nickel/zirconia interface. The nickel is out of focus due to a step created on fracturing. The micrograph shows an additional surface species at the nickel–zirconia boundary believed to be caused by $Ni(NO_3)_2$ decomposition. Figure 5(b) is the same area showing anode contact at the interface. The nickel is in good contact with the zirconia due to the presence of the fine-particle layer at the interface. The exact nature and composition of the $Ni(NO_3)_2$ treatment upon the zirconia surface are not known. Upon firing, the salt may decompose to give a layer of sub-micron

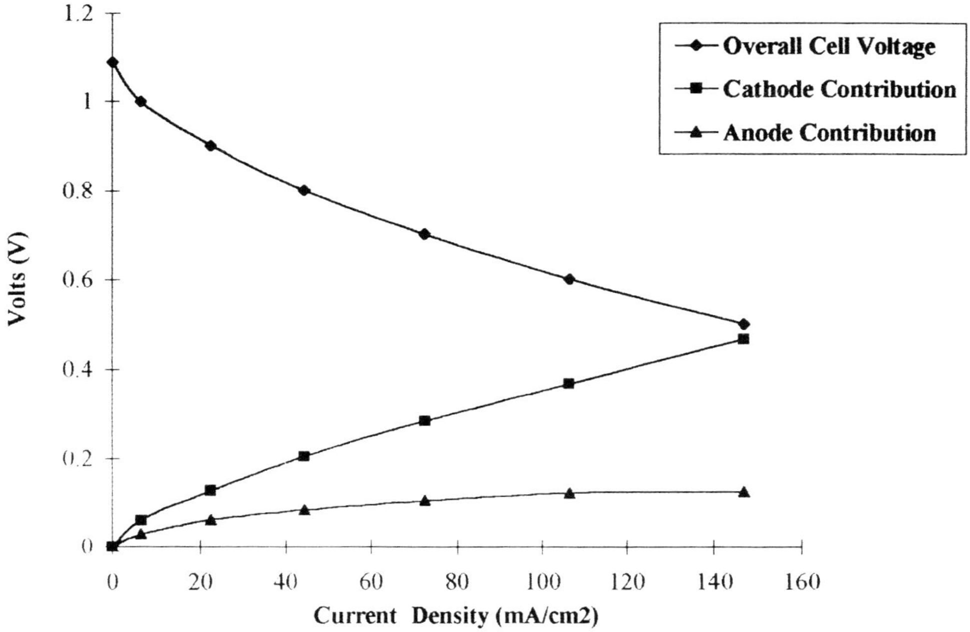

Figure 3(a) IV plot for $Ni(NO_3)_2$ and $Mn(NO_3)_2$-treated cells showing electrode contributions.

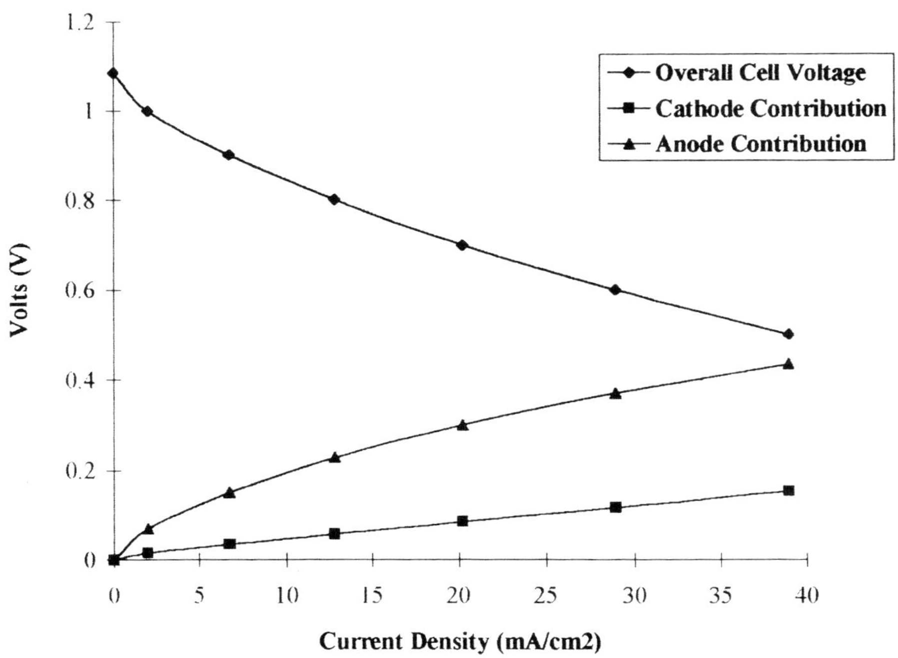

Figure 3(b) IV plot for untreated cells showing electrode contributions.

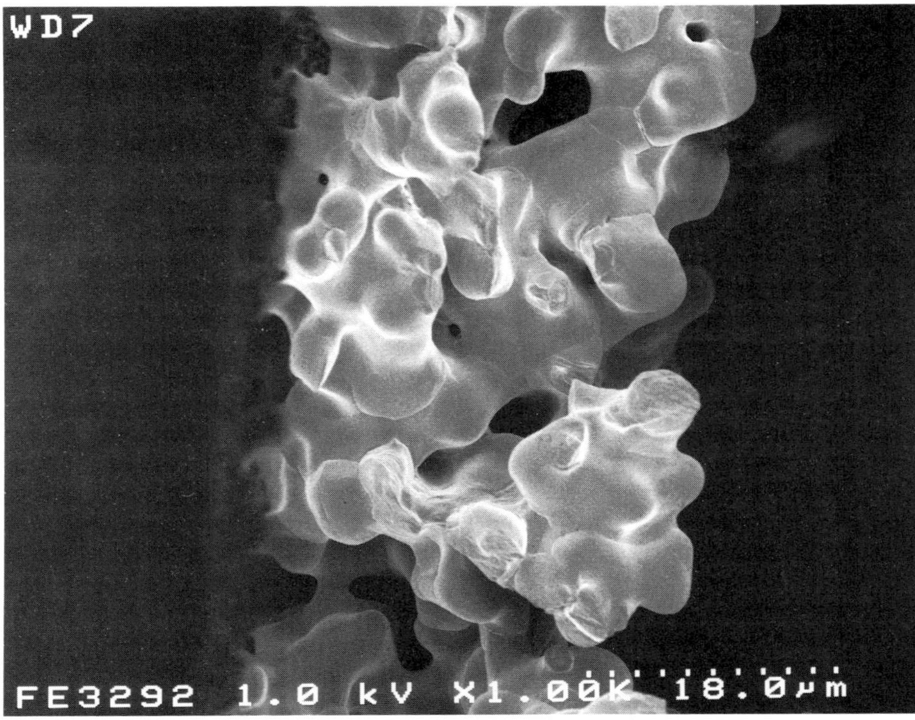

Figure 4 Low magnification SEM micrograph of untreated tube showing nickel peeling.

Table 1
Electrode resistances at 0.7 volts (cell voltage)

	Current (mA cm^{-2})	Cathode R (Ω cm^{-2})	Anode R (Ω cm^{-2})
Ni nitrate	37	7.7	3.0
Both nitrates	72	3.9	1.4
Mn nitrate	25	6.8	8.8
Untreated	20	4.2	14.9

NiO particles. These particles, upon reduction in hydrogen, allow the larger particles of the nickel anode to adhere. In addition, Fig. 5(b) shows that little nickel particle coarsening has occurred, one potential advantage of not having to fire the anode at higher temperatures.

4. CONCLUSION

The addition of nitrate solutions to the surface of zirconia increases overall cell performance when used with a nickel anode and LSM cathode. This was attributed

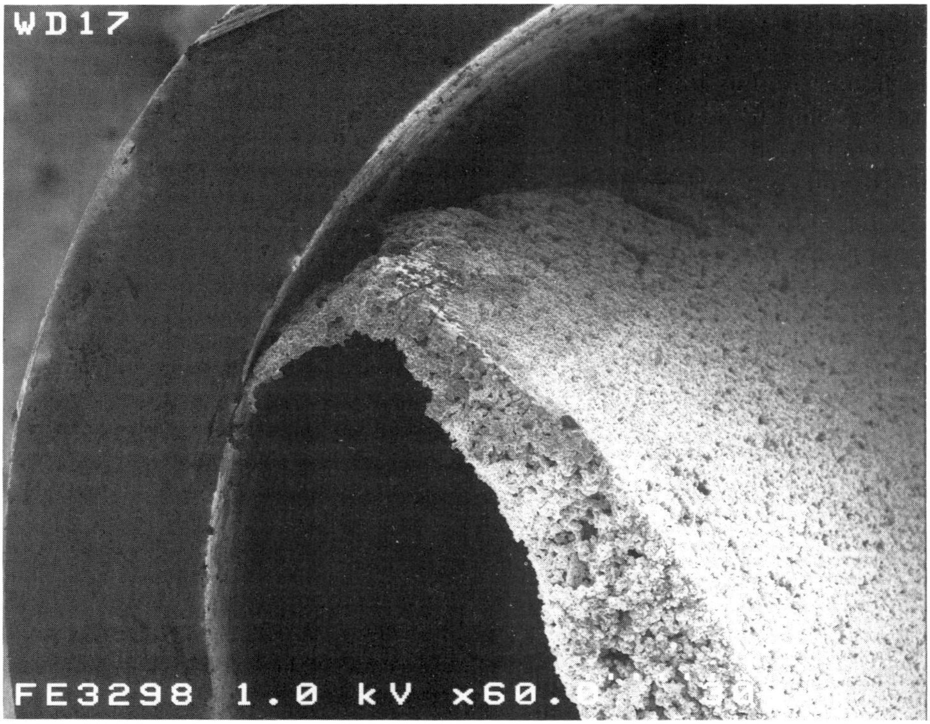

Figure 5(a) SEM micrograph of $Ni(NO_3)_2$-treated tube showing zirconia/nickel interface. Zirconia side of interface.

to improved adhesion between the nickel anode and the zirconia surface. Treating with both manganese nitrate and nickel nitrate realised the greatest improvement in performance. Additional work is needed to understand the nature of bonding between the nitrate-treated YSZ surface and the Ni anode.

ACKNOWLEDGEMENTS

The authors wish to thank EPSRC and the Brite-Euram programme for funding this work.

REFERENCES

1. K. Kendall and M. Prica: 'Integrated SOFC tubular system for small-scale cogeneration', *Proc. 1st European SOFC Forum*, 163–170.
2. A. J. McEvoy, K. R. Thampi and J. Van herle: 'Interface microstructure and electrocatalysis in SOFC', *Proc. 3rd International Symp. on SOFC*, 623–630.
3. K. R. Thampi, A. J. McEvoy and J. Van herle: 'Electrocatalysis in solid oxide fuel cell domains', *J. Electrochem. Soc.*, 1995, **142**(2), 506–513.

Figure 5(b) SEM micrograph of $Ni(NO_3)_2$-treated tube showing zirconia/nickel interface. Nickel side of interface.

4. H. Yokohama, T. Horita, N. Sakai, T. Kawada and M. Dokiya: 'Thermodynamic considerations on plausible effect of oxygen flow on morphology of reaction products at the YSZ/lanthanum manganite interface in SOFC', *Proc. 1st European SOFC Forum*, 425–434.

Author index

Subject index